Student Resources

INCLUDES
- Program Authors
- Table of Contents
- Glossary
- Common Core State Standards Correlation
- Index
- Table of Measures

Made in the United States
Text printed on 100% recycled paper

Houghton Mifflin Harcourt

Copyright © by Houghton Mifflin Harcourt Publishing Company

All rights reserved. No part of this work may be reproduced or transmitted in any form or by any means, electronic or mechanical, including photocopying or recording, or by any information storage or retrieval system, without the prior written permission of the copyright owner unless such copying is expressly permitted by federal copyright law.

Permission is hereby granted to individuals using the corresponding student's textbook or kit as the major vehicle for regular classroom instruction to photocopy entire pages from this publication in classroom quantities for instructional use and not for resale. Requests for information on other matters regarding duplication of this work should be addressed to Houghton Mifflin Harcourt Publishing Company, Attn: Contracts, Copyrights, and Licensing, 9400 Southpark Center Loop, Orlando, Florida 32819-8647.

Common Core State Standards © Copyright 2010. National Governors Association Center for Best Practices and Council of Chief State School Officers. All rights reserved.

This product is not sponsored or endorsed by the Common Core State Standards Initiative of the National Governors Association Center for Best Practices and the Council of Chief State School Officers.

Printed in the U.S.A.

ISBN 978-0-544-34347-4

16 0928 22 21 20 19 10

4500743017   B C D E F G

> If you have received these materials as examination copies free of charge, Houghton Mifflin Harcourt Publishing Company retains title to the materials and they may not be resold. Resale of examination copies is strictly prohibited.

> Possession of this publication in print format does not entitle users to convert this publication, or any portion of it, into electronic format.

GO MATH!

Authors

Juli K. Dixon, Ph.D.
Professor, Mathematics Education
University of Central Florida
Orlando, Florida

Edward B. Burger, Ph.D.
President, Southwestern University
Georgetown, Texas

Steven J. Leinwand
Principal Research Analyst
American Institutes for
 Research (AIR)
Washington, D.C.

Contributor

Rena Petrello
Professor, Mathematics
Moorpark College
Moorpark, CA

Matthew R. Larson, Ph.D.
K-12 Curriculum Specialist for
 Mathematics
Lincoln Public Schools
Lincoln, Nebraska

Martha E. Sandoval-Martinez
Math Instructor
El Camino College
Torrance, California

English Language Learners Consultant

Elizabeth Jiménez
CEO, GEMAS Consulting
Professional Expert on English
 Learner Education
Bilingual Education and
 Dual Language
Pomona, California

Table of Contents

Student Edition Table of Contents. v
Glossary H1
Common Core State Standards Correlation H14
Index H22
Table of Measures H38

Place Value and Operations with Whole Numbers

 Critical Area Developing understanding and fluency with multi-digit multiplication, and developing understanding of dividing to find quotients involving multi-digit dividends

 Food in Space 2

1 Place Value, Addition, and Subtraction to One Million — 3

Domain Number and Operations in Base Ten
COMMON CORE STATE STANDARDS 4.NBT.A.1, 4.NBT.A.2, 4.NBT.A.3, 4.NBT.B.4

✓ **Show What You Know** 3
   Vocabulary Builder 4
   Chapter Vocabulary Cards
   Vocabulary Game 4A
1 Model Place Value Relationships 5
2 Read and Write Numbers 11
3 Compare and Order Numbers 17
4 Round Numbers 23
✓ **Mid-Chapter Checkpoint** 29
5 **Investigate** • Rename Numbers 31
6 Add Whole Numbers 37
7 Subtract Whole Numbers 43
8 **Problem Solving** •
   Comparison Problems with Addition and Subtraction 49
✓ **Chapter 1 Review/Test** 55

GO DIGITAL
Go online! Your math lessons are interactive. Use iTools, Animated Math Models, the Multimedia eGlossary, and more.

Chapter 1 Overview
In this chapter, you will explore and discover answers to the following **Essential Questions**:

- How can you use place value to compare, add, subtract, and estimate with whole numbers?
- How do you compare and order whole numbers?
- What are some strategies you can use to round whole numbers?
- How is adding and subtracting 5- and 6-digit numbers similar to adding and subtracting 3-digit numbers?

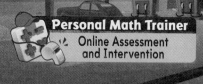

Personal Math Trainer
Online Assessment and Intervention

Chapter 2 Overview

In this chapter, you will explore and discover answers to the following **Essential Questions**:

- What strategies can you use to multiply by 1-digit numbers?
- How can you use models to multiply a multi-digit number by a 1-digit number?
- How can you use estimation to check your answer?
- How does the partial products strategy use place value?

Practice and Homework

Lesson Check and Spiral Review in every lesson

Chapter 3 Overview

In this chapter, you will explore and discover answers to the following **Essential Questions**:

- What strategies can you use to multiply 2-digit numbers?
- How can you use place value to multiply 2-digit numbers?
- How can you choose the best method to multiply 2-digit numbers?

2 Multiply by 1-Digit Numbers — 61

Domains Operations and Algebraic Thinking
Number and Operations in Base Ten

COMMON CORE STATE STANDARDS 4.OA.A.1, 4.OA.A.2, 4.OA.A.3, 4.NBT.B.5

✓ Show What You Know 61
   Vocabulary Builder 62
   Chapter Vocabulary Cards
   Vocabulary Game 62A
1 Algebra • Multiplication Comparisons 63
2 Algebra • Comparison Problems 69
3 Multiply Tens, Hundreds, and Thousands 75
4 Estimate Products 81
5 Investigate • Multiply Using the Distributive Property . . . 87
6 Multiply Using Expanded Form 93
7 Multiply Using Partial Products 99
✓ Mid-Chapter Checkpoint 105
8 Multiply Using Mental Math 107
9 Problem Solving • Multistep Multiplication Problems . . . 113
10 Multiply 2-Digit Numbers with Regrouping 119
11 Multiply 3-Digit and 4-Digit Numbers with Regrouping . . 125
12 Algebra • Solve Multistep Problems Using Equations . . . 131
✓ Chapter 2 Review/Test 137

3 Multiply 2-Digit Numbers — 143

Domains Operations and Algebraic Thinking
Number and Operations in Base Ten

COMMON CORE STATE STANDARDS 4.OA.A.3, 4.NBT.B.5

✓ Show What You Know 143
   Vocabulary Builder 144
   Chapter Vocabulary Cards
   Vocabulary Game 144A
1 Multiply by Tens 145
2 Estimate Products 151
3 Investigate • Area Models and Partial Products 157
4 Multiply Using Partial Products 163
✓ Mid-Chapter Checkpoint 169
5 Multiply with Regrouping 171
6 Choose a Multiplication Method 177
7 Problem Solving • Multiply 2-Digit Numbers 183
✓ Chapter 3 Review/Test 189

Divide by 1-Digit Numbers — 195

Domains Operations and Algebraic Thinking
Number and Operations in Base Ten
COMMON CORE STATE STANDARDS 4.OA.A.3, 4.NBT.B.6

✓ Show What You Know 195
Vocabulary Builder 196
Chapter Vocabulary Cards
Vocabulary Game 196A
1 Estimate Quotients Using Multiples 197
2 Investigate • Remainders 203
3 Interpret the Remainder 209
4 Divide Tens, Hundreds, and Thousands 215
5 Estimate Quotients Using Compatible Numbers 221
6 Investigate • Division and the Distributive Property 227
✓ Mid-Chapter Checkpoint 233
7 Investigate • Divide Using Repeated Subtraction 235
8 Divide Using Partial Quotients 241
9 Investigate • Model Division with Regrouping 247
10 Place the First Digit 253
11 Divide by 1-Digit Numbers 259
12 Problem Solving • Multistep Division Problems 265
✓ Chapter 4 Review/Test 271

Chapter 4 Overview

In this chapter, you will explore and discover answers to the following **Essential Questions**:

- How can you divide by 1-digit numbers?
- How can you use remainders in division problems?
- How can you estimate quotients?
- How can you model division with a 1-digit divisor?

Chapter 5 Overview

In this chapter, you will explore and discover answers to the following **Essential Questions**:

- How can you find factors and multiples, and how can you generate and describe number patterns?
- How can you use models or lists to find factors?
- How can you create a number pattern?

5 Factors, Multiples, and Patterns 277

Domain Operations and Algebraic Thinking
COMMON CORE STATE STANDARDS 4.OA.B.4, 4.OA.C.5

✓ Show What You Know 277
Vocabulary Builder 278
Chapter Vocabulary Cards
Vocabulary Game 278A
1 Model Factors 279
2 Factors and Divisibility 285
3 Problem Solving • Common Factors 291
✓ Mid-Chapter Checkpoint 297
4 Factors and Multiples 299
5 Prime and Composite Numbers 305
6 Algebra • Number Patterns 311
✓ Chapter 5 Review/Test 317

Fractions and Decimals

 Critical Area Developing an understanding of fraction equivalence, addition and subtraction of fractions with like denominators, and multiplication of fractions by whole numbers

Real World Project Building Custom Guitars **324**

6 Fraction Equivalence and Comparison — 325

Domain Number and Operations–Fractions
COMMON CORE STATE STANDARDS 4.NF.A.1, 4.NF.A.2

- ✓ Show What You Know **325**
- Vocabulary Builder **326**
- Chapter Vocabulary Cards
- Vocabulary Game **326A**
- 1 Investigate • Equivalent Fractions **327**
- 2 Generate Equivalent Fractions **333**
- 3 Simplest Form **339**
- 4 Common Denominators **345**
- 5 Problem Solving • Find Equivalent Fractions **351**
- ✓ Mid-Chapter Checkpoint **357**
- 6 Compare Fractions Using Benchmarks **359**
- 7 Compare Fractions **365**
- 8 Compare and Order Fractions **371**
- ✓ Chapter 6 Review/Test **377**

7 Add and Subtract Fractions — 383

Domain Number and Operations–Fractions
COMMON CORE STATE STANDARDS 4.NF.B.3a, 4.NF.B.3b, 4.NF.B.3c, 4.NF.B.3d

- ✓ Show What You Know **383**
- Vocabulary Builder **384**
- Chapter Vocabulary Cards
- Vocabulary Game **384A**
- 1 Investigate • Add and Subtract Parts of a Whole **385**
- 2 Write Fractions as Sums **391**
- 3 Add Fractions Using Models **397**
- 4 Subtract Fractions Using Models **403**
- 5 Add and Subtract Fractions **409**
- ✓ Mid-Chapter Checkpoint **415**
- 6 Rename Fractions and Mixed Numbers **417**
- 7 Add and Subtract Mixed Numbers **423**
- 8 Subtraction with Renaming **429**
- 9 Algebra • Fractions and Properties of Addition **435**
- 10 Problem Solving • Multistep Fraction Problems **441**
- ✓ Chapter 7 Review/Test **447**

Critical Area

GO DIGITAL
Go online! Your math lessons are interactive. Use iTools, Animated Math Models, the Multimedia eGlossary, and more.

Chapter 6 Overview
Essential Questions:
- What strategies can you use to compare fractions and write equivalent fractions?
- What models can help you compare and order fractions?
- How can you find equivalent fractions?
- How can you solve problems that involve fractions?

Chapter 7 Overview
Essential Questions:
- How do you add or subtract fractions that have the same denominator?
- Why do you add or subtract the numerators and not the denominators?
- Why do you rename mixed numbers when adding or subtracting fractions?
- How do you know that your sum or difference is reasonable?

Chapter 8 Overview

In this chapter, you will explore and discover answers to the following **Essential Questions**:

- How do you multiply fractions by whole numbers?
- How can you write a product of a whole number and a fraction as a product of a whole number and a unit fraction?

Practice and Homework

Lesson Check and Spiral Review in every lesson

8 Multiply Fractions by Whole Numbers 453

Domain Number and Operations–Fractions
COMMON CORE STATE STANDARDS 4.NF.B.4a, 4.NF.B.4b, 4.NF.B.4c

- ✓ Show What You Know 453
- Vocabulary Builder 454
- Chapter Vocabulary Cards
- Vocabulary Game 454A
- 1 Multiples of Unit Fractions. 455
- 2 Multiples of Fractions. 461
- ✓ Mid-Chapter Checkpoint 467
- 3 Multiply a Fraction by a Whole Number Using Models 469
- 4 Multiply a Fraction or Mixed Number by a Whole Number. 475
- 5 Problem Solving •
  Comparison Problems with Fractions. 481
- ✓ Chapter 8 Review/Test 487

Chapter 9 Overview

In this chapter, you will explore and discover answers to the following **Essential Questions**:

- How can you record decimal notation for fractions and compare decimal fractions?
- Why can you record tenths and hundredths as decimals and fractions?
- What are some different models you can use to find equivalent fractions?
- How can you compare decimal fractions?

9 Relate Fractions and Decimals 493

Domains Number and Operations–Fractions
  Measurement and Data
COMMON CORE STATE STANDARDS 4.NF.C.5, 4.NF.C.6, 4.NF.C.7, 4.MD.A.2

- ✓ Show What You Know 493
- Vocabulary Builder 494
- Chapter Vocabulary Cards
- Vocabulary Game 494A
- 1 Relate Tenths and Decimals 495
- 2 Relate Hundredths and Decimals 501
- 3 Equivalent Fractions and Decimals 507
- 4 Relate Fractions, Decimals, and Money 513
- 5 Problem Solving • Money 519
- ✓ Mid-Chapter Checkpoint 525
- 6 Add Fractional Parts of 10 and 100 527
- 7 Compare Decimals 533
- ✓ Chapter 9 Review/Test 539

Geometry, Measurement, and Data

 Critical Area Understanding that geometric figures can be analyzed and classified based on their properties, such as having parallel sides, perpendicular sides, particular angle measures, and symmetry

Real World Project Landscape Architects 546

10 Two-Dimensional Figures 547

Domains Operations and Algebraic Thinking
 Geometry
COMMON CORE STATE STANDARDS 4.OA.C.5, 4.G.A.1, 4.G.A.2, 4.G.A.3

- ✓ Show What You Know 547
- Vocabulary Builder 548
- Chapter Vocabulary Cards
- Vocabulary Game 548A
- **1** Lines, Rays, and Angles 549
- **2** Classify Triangles by Angles 555
- **3** Parallel Lines and Perpendicular Lines 561
- **4** Classify Quadrilaterals 567
- ✓ Mid-Chapter Checkpoint 573
- **5** Line Symmetry 575
- **6** Find and Draw Lines of Symmetry 581
- **7** Problem Solving • Shape Patterns 587
- ✓ Chapter 10 Review/Test 593

Chapter 10 Overview
Essential Questions:
- How can you draw and identify lines and angles, and how can you classify shapes?
- What are the building blocks of geometry?
- How can you classify triangles and quadrilaterals?
- How do you recognize symmetry in a polygon?

11 Angles 599

Domain Measurement and Data
COMMON CORE STATE STANDARDS 4.MD.C.5a, 4.MD.C.5b, 4.MD.C.6, 4.MD.C.7

- ✓ Show What You Know 599
- Vocabulary Builder 600
- Chapter Vocabulary Cards
- Vocabulary Game 600A
- **1** Investigate • Angles and Fractional Parts of a Circle 601
- **2** Degrees 607
- **3** Measure and Draw Angles 613
- ✓ Mid-Chapter Checkpoint 619
- **4** Investigate • Join and Separate Angles 621
- **5** Problem Solving • Unknown Angle Measures 627
- ✓ Chapter 11 Review/Test 633

Chapter 11 Overview
Essential Questions:
- How can you measure angles and solve problems involving angle measures?
- How can you use fractions and degrees to understand angle measures?
- How can you use a protractor to measure and classify angles?
- How can equations help you find the measurement of an angle?

xi

Chapter 12 Overview

In this chapter, you will explore and discover answers to the following **Essential Questions**:

- How can you use relative sizes of measurements to solve problems and to generate measurement tables that show a relationship?
- How can you compare metric units of length, mass, or liquid volume?
- How can you compare customary units of length, weight, or liquid volume?

Practice and Homework

Lesson Check and Spiral Review in every lesson

12 Relative Sizes of Measurement Units — 639

Domain Measurement and Data
COMMON CORE STATE STANDARDS 4.MD.A.1, 4.MD.A.2, 4.MD.B.4

✓ Show What You Know 639
  Vocabulary Builder 640
  Chapter Vocabulary Cards
  Vocabulary Game 640A
1 Measurement Benchmarks 641
2 Customary Units of Length 647
3 Customary Units of Weight 653
4 Customary Units of Liquid Volume 659
5 Line Plots 665
✓ Mid-Chapter Checkpoint 671
6 Investigate • Metric Units of Length 673
7 Metric Units of Mass and Liquid Volume 679
8 Units of Time 685
9 Problem Solving • Elapsed Time 691
10 Mixed Measures 697
11 Algebra • Patterns in Measurement Units 703
✓ Chapter 12 Review/Test 709

Chapter 13 Overview

In this chapter, you will explore and discover answers to the following **Essential Questions**:

- How can you use formulas for perimeter and area to solve problems?
- How are area and perimeter different?
- What are some methods you can use to find area and perimeter of a figure?
- How could two different rectangles have the same perimeter or the same area?

13 Algebra: Perimeter and Area — 715

Domain Measurement and Data
COMMON CORE STATE STANDARDS 4.MD.A.3

✓ Show What You Know 715
  Vocabulary Builder 716
  Chapter Vocabulary Cards
  Vocabulary Game 716A
1 Perimeter 717
2 Area 723
3 Area of Combined Rectangles 729
✓ Mid-Chapter Checkpoint 735
4 Find Unknown Measures 737
5 Problem Solving • Find the Area 743
✓ Chapter 13 Review/Test 749

Glossary H1
Common Core State Standards H14
Index H22
Table of Measures H38

xii

Glossary

Pronunciation Key

a add, map	ē equal, tree	m move, seem	o͞o pool, food	u̇ pull, book
ā ace, rate	f fit, half	n nice, tin	p pit, stop	û(r) burn, term
â(r) care, air	g go, log	ng ring, song	r run, poor	yo͞o fuse, few
ä palm, father	h hope, hate	o odd, hot	s see, pass	v vain, eve
	i it, give	ō open, so	sh sure, rush	w win, away
b bat, rub	ī ice, write	ô order, jaw	t talk, sit	y yet, yearn
ch check, catch	j joy, ledge	oi oil, boy	th thin, both	z zest, muse
d dog, rod	k cool, take	ou pout, now	th this, bathe	zh vision, pleasure
e end, pet	l look, rule	o͝o took, full	u up, done	

ə the schwa, an unstressed vowel representing the sound spelled *a* in *above*, *e* in *sicken*, *i* in *possible*, *o* in *melon*, *u* in *circus*

Other symbols:
• separates words into syllables
′ indicates stress on a syllable

acute angle [ə•kyo͞ot′ ang′gəl] **ángulo agudo**
An angle that measures greater than 0° and less than 90°
Example:

acute triangle [ə•kyo͞ot′ trī′ang•gəl]
triángulo acutángulo A triangle with three acute angles
Example:

addend [a′dend] **sumando** A number that is added to another in an addition problem
Example: 2 + 4 = 6;
2 and 4 are addends.

addition [ə•di′shən] **suma** The process of finding the total number of items when two or more groups of items are joined; the opposite operation of subtraction

A.M. [ā•em′] **a.m.** The times after midnight and before noon

analog clock [anəl• ôg kläk] **reloj analógico**
A tool for measuring time, in which hands move around a circle to show hours, minutes, and sometimes seconds
Example:

angle [ang′gəl] **ángulo** A shape formed by two line segments or rays that share the same endpoint
Example:

area [âr′ē•ə] **área** The measure of the number of unit squares needed to cover a surface
Example:

Area = 9 square units

Glossary **H1**

array [ə•rā′] **matriz** An arrangement of objects in rows and columns
Example:

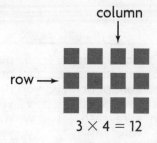

$3 \times 4 = 12$

Associative Property of Addition [ə•sō′shē•āt•iv präp′ər•tē əv ə•dish′ən] **propiedad asociativa de la suma** The property that states that you can group addends in different ways and still get the same sum
Example: $3 + (8 + 5) = (3 + 8) + 5$

Associative Property of Multiplication [ə•sō′shē•ə•tiv präp′ər•tē əv mul•tə•pli•kā′shən] **propiedad asociativa de la multiplicación** The property that states that you can group factors in different ways and still get the same product
Example: $3 \times (4 \times 2) = (3 \times 4) \times 2$

bar graph [bär graf] **gráfica de barras** A graph that uses bars to show data
Example:

base [bās] **base** A polygon's side or a two-dimensional shape, usually a polygon or circle, by which a three-dimensional shape is measured or named
Examples:

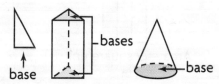

benchmark [bench′märk] **punto de referencia** A known size or amount that helps you understand a different size or amount

calendar [kal′ən•dər] **calendario** A table that shows the days, weeks, and months of a year

capacity [kə•pas′i•tē] **capacidad** The amount a container can hold when filled

Celsius (°C) [sel′sē•əs] **Celsius** A metric scale for measuring temperature

centimeter (cm) [sen′tə•mēt•ər] **centímetro (cm)** A metric unit for measuring length or distance
1 meter = 100 centimeters
Example:

1 centimeter

cent sign (¢) [sent sīn] **símbolo de centavo** A symbol that stands for *cent* or *cents*
Example: 53¢

clockwise [kläk′wīz] **en el sentido de las manecillas del reloj** In the same direction in which the hands of a clock move

closed shape [klōzd shāp] **figura cerrada** A two-dimensional shape that begins and ends at the same point
Examples:

common denominator [käm′ən dē•näm′ə•nāt•ər] **denominador común** A common multiple of two or more denominators
Example: Some common denominators for $\frac{1}{4}$ and $\frac{5}{6}$ are 12, 24, and 36.

common factor [käm′ən fak′tər] **factor común** A number that is a factor of two or more numbers

common multiple [käm′ən mul′tə•pəl] **múltiplo común** A number that is a multiple of two or more numbers

H2 **Glossary**

Commutative Property of Addition
[kə•myo͞ot′ə•tiv präp′ər•tē əv ə•dish′ən] **propiedad conmutativa de la suma** The property that states that when the order of two addends is changed, the sum is the same
Example: $4 + 5 = 5 + 4$

Commutative Property of Multiplication
[kə•myo͞ot′ə•tiv präp′ər•tē əv mul•tə•pli•kā′shən] **propiedad conmutativa de la multiplicación** The property that states that when the order of two factors is changed, the product is the same
Example: $4 \times 5 = 5 \times 4$

compare [kəm•pâr′] **comparar** To describe whether numbers are equal to, less than, or greater than each other

compatible numbers [kəm•pat′ə•bəl num′bərz] **números compatibles** Numbers that are easy to compute mentally

composite number [kəm•päz′it num′bər] **número compuesto** A number having more than two factors
Example: 6 is a composite number, since its factors are 1, 2, 3, and 6.

corner [kôr′nər] **esquina** See *vertex*.

counterclockwise [kount•er•kläk′wīz] **en sentido contrario a las manecillas del reloj** In the opposite direction in which the hands of a clock move

counting number [kount′ing num′bər] **número natural** A whole number that can be used to count a set of objects (1, 2, 3, 4, . . .)

cube [kyo͞ob] **cubo** A three-dimensional shape with six square faces of the same size
Example:

cup (c) [kup] **taza (tz)** A customary unit used to measure capacity and liquid volume
1 cup = 8 ounces

data [dāt′ə] **datos** Information collected about people or things

decagon [dek′ə•gän] **decágono** A polygon with ten sides and ten angles

decimal [des′ə•məl] **decimal** A number with one or more digits to the right of the decimal point

decimal point [des′ə•məl point] **punto decimal** A symbol used to separate dollars from cents in money amounts, and to separate the ones and the tenths places in a decimal
Example: 6.4
↑ decimal point

decimeter (dm) [des′i•mēt•ər] **decímetro (dm)** A metric unit for measuring length or distance
1 meter = 10 decimeters

degree (°) [di•grē′] **grado (°)** The unit used for measuring angles and temperatures

denominator [dē•näm′ə•nāt•ər] **denominador** The number below the bar in a fraction that tells how many equal parts are in the whole or in the group
Example: $\frac{3}{4}$ ← denominator

diagonal [dī•ag′ə•nəl] **diagonal** A line segment that connects two vertices of a polygon that are not next to each other
Example:

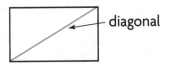

difference [dif′ər•əns] **diferencia** The answer to a subtraction problem

digit [dij′it] **dígito** Any one of the ten symbols 0, 1, 2, 3, 4, 5, 6, 7, 8, or 9 used to write numbers

digital clock [dij′i•təl kläk] **reloj digital** A clock that shows time to the minute, using digits
Example:

Glossary H3

dime [dīm] **moneda de 10¢** A coin worth 10 cents and with a value equal to that of 10 pennies; 10¢
Example:

dimension [də•men'shən] **dimensión** A measure in one direction

Distributive Property [di•strib'yoō•tiv präp'ər•tē] **propiedad distributiva** The property that states that multiplying a sum by a number is the same as multiplying each addend by the number and then adding the products
Example: $5 \times (10 + 6) = (5 \times 10) + (5 \times 6)$

divide [də•vīd'] **dividir** To separate into equal groups; the opposite operation of multiplication

dividend [dəv'ə•dend] **dividendo** The number that is to be divided in a division problem
Example: $36 \div 6$; $6\overline{)36}$; the dividend is 36.

divisible [də•viz'ə•bəl] **divisible** A number is divisible by another number if the quotient is a counting number and the remainder is zero
Example: 18 is divisible by 3.

division [də•vi'zhən] **división** The process of sharing a number of items to find how many equal groups can be made or how many items will be in each equal group; the opposite operation of multiplication

divisor [də•vī'zər] **divisor** The number that divides the dividend
Example: $15 \div 3$; $3\overline{)15}$; the divisor is 3.

dollar [däl'ər] **dólar** Paper money worth 100 cents and equal to 100 pennies; $1.00
Example:

elapsed time [ē•lapst' tīm] **tiempo transcurrido** The time that passes from the start of an activity to the end of that activity

endpoint [end'point] **extremo** The point at either end of a line segment or the starting point of a ray

equal groups [ē'kwəl groōpz] **grupos iguales** Groups that have the same number of objects

equal parts [ē'kwəl pärts] **partes iguales** Parts that are exactly the same size

equal sign (=) [ē'kwəl sīn] **signo de igualdad** A symbol used to show that two numbers have the same value
Example: $384 = 384$

equal to [ē'kwəl toō] **igual a** Having the same value
Example: $4 + 4$ is equal to $3 + 5$.

equation [ē•kwā'zhən] **ecuación** A number sentence which shows that two quantities are equal
Example: $4 + 5 = 9$

equivalent [ē•kwiv'ə•lənt] **equivalente** Having the same value or naming the same amount

equivalent decimals [ē•kwiv'ə•lənt des'ə•məlz] **decimales equivalentes** Two or more decimals that name the same amount

equivalent fractions [ē•kwiv'ə•lənt frak'shənz] **fracciones equivalentes** Two or more fractions that name the same amount
Example: $\frac{3}{4}$ and $\frac{6}{8}$ name the same amount.

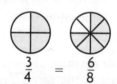

$$\frac{3}{4} = \frac{6}{8}$$

estimate [es'tə•māt] *verb* **estimar** To find an answer that is close to the exact amount

estimate [es'tə•mit] *noun* **estimación** A number that is close to the exact amount

even [ē'vən] **par** A whole number that has a 0, 2, 4, 6, or 8 in the ones place

expanded form [ek•span'did fôrm] **forma desarrollada** A way to write numbers by showing the value of each digit
Example: $253 = 200 + 50 + 3$

expression [ek•spresh'ən] **expresión** A part of a number sentence that has numbers and operation signs but does not have an equal sign

fact family [fakt fam′ə•lē] **familia de operaciones** A set of related multiplication and division equations, or addition and subtraction equations
Example: $7 \times 8 = 56 \quad 8 \times 7 = 56$
$56 \div 7 = 8 \quad 56 \div 8 = 7$

factor [fak′tər] **factor** A number that is multiplied by another number to find a product

Fahrenheit (°F) [fâr′ən•hīt] **Fahrenheit** A customary scale for measuring temperature

fluid ounce (fl oz) [flōō′id ouns] **onza fluida (fl oz)** A customary unit used to measure liquid capacity and liquid volume
1 cup = 8 fluid ounces

foot (ft) [fŏŏt] **pie (ft)** A customary unit used for measuring length or distance
1 foot = 12 inches

formula [fôr′myōō•lə] **fórmula** A set of symbols that expresses a mathematical rule
Example: Area = base × height, or $A = b \times h$

fraction [frak′shən] **fracción** A number that names a part of a whole or part of a group
Example:

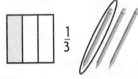

fraction greater than 1 [frak′shən grāt′ər than wun] **fracción mayor que 1** A number which has a numerator that is greater than its denominator

frequency table [frē′kwən•sē tā′bəl] **tabla de frecuencia** A table that uses numbers to record data about how often something happens
Example:

Favorite Color	
Color	Frequency
Blue	10
Red	7
Green	5
Other	3

gallon (gal) [gal′ən] **galón (gal)** A customary unit for measuring capacity and liquid volume
1 gallon = 4 quarts

gram (g) [gram] **gramo (g)** A metric unit for measuring mass
1 kilogram = 1,000 grams

greater than sign (>) [grāt′ər than sīn] **signo de mayor que** A symbol used to compare two quantities, with the greater quantity given first
Example: $6 > 4$

grid [grid] **cuadrícula** Evenly divided and equally spaced squares on a shape or flat surface

half gallon [haf gal′ən] **medio galón** A customary unit for measuring capacity and liquid volume
1 half gallon = 2 quarts

half hour [haf our] **media hora** 30 minutes
Example: 4:00 to 4:30 is one half hour.

half-square unit [haf skwâr yōō′nit] **media unidad cuadrada** Half of a unit of area with dimensions of 1 unit × 1 unit

height [hīt] **altura** The measure of a perpendicular from the base to the top of a two-dimensional shape

hexagon [hek′sə•gän] **hexágono** A polygon with six sides and six angles
Examples:

horizontal [hôr•i•zänt′l] **horizontal** In the direction from left to right

hour (hr) [our] **hora (hr)** A unit used to measure time
1 hour = 60 minutes

hundredth [hun′drədth] **centésimo** One of one hundred equal parts
Example:

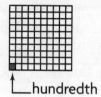
hundredth

Identity Property of Addition [ī•den′tə•tē präp′ər•tē əv ə•dish′ən] **propiedad de identidad de la suma** The property that states that when you add zero to any number, the sum is that number
Example: 16 + 0 = 16

Identity Property of Multiplication [ī•den′tə•tē präp′ər•tē əv mul•tə•pli•kā′shən] **propiedad de identidad de la multiplicación** The property that states that the product of any number and 1 is that number
Example: 9 × 1 = 9

inch (in.) [inch] **pulgada (pulg)** A customary unit used for measuring length or distance
Example:

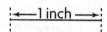

intersecting lines [in•tər•sekt′ing līnz] **líneas secantes** Lines that cross each other at exactly one point
Example:

inverse operations [in′vûrs äp•ə•rā′shənz] **operaciones inversas** Operations that undo each other, such as addition and subtraction or multiplication and division
Example: 6 × 8 = 48 and 48 ÷ 6 = 8

key [kē] **clave** The part of a map or graph that explains the symbols

kilogram (kg) [kil′ō•gram] **kilogramo (kg)** A metric unit for measuring mass
1 kilogram = 1,000 grams

kilometer (km) [kə•läm′ət•ər] **kilómetro (km)** A metric unit for measuring length or distance
1 kilometer = 1,000 meters

length [lengkth] **longitud** The measurement of the distance between two points

less than sign (<) [les than sīn] **signo de menor que** A symbol used to compare two quantities, with the lesser quantity given first
Example: 3 < 7

line [līn] **línea** A straight path of points in a plane that continues without end in both directions with no endpoints
Example:

S   T

line graph [līn graf] **gráfica lineal** A graph that uses line segments to show how data change over time

line of symmetry [līn əv sim′ə•trē] **eje de simetría** An imaginary line on a shape about which the shape can be folded so that its two parts match exactly
Example:

line of symmetry

line plot [līn plöt] **diagrama de puntos** A graph that records each piece of data on a number line
Example:

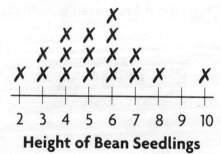
Height of Bean Seedlings

H6 Glossary

line segment [līn seg′mənt] **segmento** A part of a line that includes two points called endpoints and all the points between them
Example:

line symmetry [līn sim′ə•trē] **simetría axial** What a shape has if it can be folded about a line so that its two parts match exactly

linear units [lin′ē•ər yōō′nits] **unidades lineales** Units that measure length, width, height, or distance

liquid volume [lik′wid väl′yōōm] **volumen de un líquido** The measure of the space a liquid occupies

liter (L) [lēt′ər] **litro (L)** A metric unit for measuring capacity and liquid volume
1 liter = 1,000 milliliters

mass [mas] **masa** The amount of matter in an object

meter (m) [mēt′ər] **metro (m)** A metric unit for measuring length or distance
1 meter = 100 centimeters

midnight [mid′nīt] **medianoche** 12:00 at night

mile (mi) [mīl] **milla (mi)** A customary unit for measuring length or distance
1 mile = 5,280 feet

milliliter (mL) [mil′i•lēt•ər] **mililitro (mL)** A metric unit for measuring capacity and liquid volume
1 liter = 1,000 milliliters

millimeter (mm) [mil′i•mēt•ər] **milímetro (mm)** A metric unit for measuring length or distance
1 centimeter = 10 millimeters

million [mil′yən] **millón** The counting number after 999,999; 1,000 thousands; written as 1,000,000

millions [mil′yənz] **millones** The period after thousands

minute (min) [min′it] **minuto (min)** A unit used to measure short amounts of time
1 minute = 60 seconds

mixed number [mikst num′bər] **número mixto** An amount given as a whole number and a fraction

multiple [mul′tə•pəl] **múltiplo** The product of a number and a counting number is called a multiple of the number
Example:

```
   3      3      3      3
 × 1    × 2    × 3    × 4    ← counting numbers
 ───    ───    ───    ───
   3      6      9     12    ← multiples of 3
```

multiplication [mul•tə•pli•kā′shən] **multiplicación** A process to find the total number of items in equal-sized groups, or to find the total number of items in a given number of groups when each group contains the same number of items; multiplication is the inverse of division

multiply [mul′tə•pli] **multiplicar** To combine equal groups to find how many in all; the opposite operation of division

nickel [nik′əl] **moneda de 5¢** A coin worth 5 cents and with a value equal to that of 5 pennies; 5¢
Example:

noon [nōōn] **mediodía** 12:00 in the day

not equal to sign (≠) [not ē′kwəl tōō sīn] **signo de no igual a** A symbol that indicates one quantity is not equal to another
Example: 12 × 3 ≠ 38

number line [num′bər līn] **recta numérica** A line on which numbers can be located
Example:

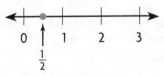

number sentence [num′bər sent′ns] **enunciado numérico** A sentence that includes numbers, operation symbols, and a greater than or less than symbol or an equal sign
Example: 5 + 3 = 8

Glossary H7

numerator [nōō′mər•āt•ər] **numerador** The number above the bar in a fraction that tells how many parts of the whole or group are being considered

Example: $\frac{2}{3}$ ← numerator

obtuse angle [äb•tōōs′ ang′gəl] **ángulo obtuso** An angle that measures greater than 90° and less than 180°
Example:

> **Word History**
>
> The Latin prefix **ob-** means "against." When combined with **-tusus**, meaning "beaten," the Latin word *obtusus*, from which we get *obtuse*, means "beaten against." This makes sense when you look at an obtuse angle, because the angle is not sharp or acute. The angle looks as if it has been beaten against and become blunt and rounded.

obtuse triangle [äb•tōōs′ trī′ang•gəl] **triángulo obtusángulo** A triangle with one obtuse angle

Example:

octagon [äk′tə•gän] **octágono** A polygon with eight sides and eight angles
Examples:

odd [od] **impar** A whole number that has a 1, 3, 5, 7, or 9 in the ones place

one-dimensional [wun də•men′shə•nəl] **unidimensional** Measured in only one direction, such as length
Examples:

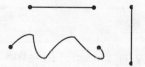

open shape [ō′pən shāp] **figura abierta** A shape that does not begin and end at the same point
Examples:

order [ôr′dər] **orden** A particular arrangement or placement of things one after the other

order of operations [ôr′dər əv äp•ə•rā′shənz] **orden de las operaciones** A special set of rules which gives the order in which calculations are done

ounce (oz) [ouns] **onza (oz)** A customary unit for measuring weight
1 pound = 16 ounces

parallel lines [pâr′ə•lel līnz] **líneas paralelas** Lines in the same plane that never intersect and are always the same distance apart
Example:

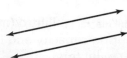

> **Word History**
>
> Euclid, an early Greek mathematician, was one of the first to explore the idea of parallel lines. The prefix **para-** means "beside or alongside." This prefix helps you understand the meaning of the word *parallel*.

parallelogram [pâr·ə·lel´ə·gram] **paralelogramo** A quadrilateral whose opposite sides are parallel and of equal length
Example:

parentheses [pə·ren´thə·sēz] **paréntesis** The symbols used to show which operation or operations in an expression should be done first

partial product [pär´shəl präd´əkt] **producto parcial** A method of multiplying in which the ones, tens, hundreds, and so on are multiplied separately and then the products are added together

partial quotient [pär´shəl kwō´shənt] **cociente parcial** A method of dividing in which multiples of the divisor are subtracted from the dividend and then the quotients are added together

pattern [pat´ərn] **patrón** An ordered set of numbers or objects; the order helps you predict what will come next
Examples: 2, 4, 6, 8, 10

pattern unit [pat´ərn yoō´nit] **unidad de patrón** The part of a pattern that repeats
Example:

pentagon [pen´tə·gän] **pentágono** A polygon with five sides and five angles
Examples:

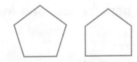

perimeter [pə·rim´ə·tər] **perímetro** The distance around a shape

period [pir´ē·əd] **período** Each group of three digits in a multi-digit number; periods are usually separated by commas or spaces.
Example: 85,643,900 has three periods.

perpendicular lines [pər·pən·dik´yoō·lər līnz] **líneas perpendiculares** Two lines that intersect to form four right angles
Example:

picture graph [pik´chər graf] **gráfica con dibujos** A graph that uses symbols to show and compare information
Example:

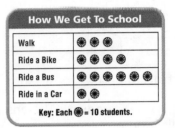

pint (pt) [pīnt] **pinta (pt)** A customary unit for measuring capacity and liquid volume
1 pint = 2 cups

place value [plās val´yoō] **valor posicional** The value of a digit in a number, based on the location of the digit

plane [plān] **plano** A flat surface that extends without end in all directions
Example:

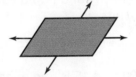

plane shape [plān shāp] **figura plana** See *two-dimensional figure.*

P.M. [pē´em] **p.m.** The times after noon and before midnight

point [point] **punto** An exact location in space

polygon [päl´i·gän] **polígono** A closed two-dimensional shape formed by three or more straight sides that are line segments
Examples:

Polygons

Not Polygons

Glossary H9

pound (lb) [pound] **libra (lb)** A customary unit for measuring weight
1 pound = 16 ounces

prime number [prīm num'bər] **número primo** A number that has exactly two factors: 1 and itself
Examples: 2, 3, 5, 7, 11, 13, 17, and 19 are prime numbers. 1 is not a prime number.

prism [priz'əm] **prisma** A solid figure that has two same size, same polygon-shaped bases, and other faces that are all rectangles
Examples:

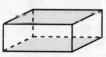

rectangular prism   triangular prism

product [präd'əkt] **producto** The answer to a multiplication problem

protractor [prō'trak•tər] **transportador** A tool for measuring the size of an angle

quadrilateral [kwä•dri•lat'ər•əl] **cuadrilátero** A polygon with four sides and four angles

quart (qt) [kwôrt] **cuarto (ct)** A customary unit for measuring capacity and liquid volume
1 quart = 2 pints

quarter hour [kwôrt'ər our] **cuarto de hora** 15 minutes
Example: 4:00 to 4:15 is one quarter hour

quotient [kwō'shənt] **cociente** The number, not including the remainder, that results from dividing
Example: 8 ÷ 4 = 2; 2 is the quotient.

ray [rā] **semirrecta** A part of a line; it has one endpoint and continues without end in one direction
Example:

K   L

rectangle [rek'tang•gəl] **rectángulo** A quadrilateral with two pairs of parallel sides, two pairs of sides of equal length, and four right angles
Example:

rectangular prism [rek•tang'gyə•lər priz'əm] **prisma rectangular** A three-dimensional shape in which all six faces are rectangles
Example:

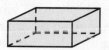

regroup [rē•grōop'] **reagrupar** To exchange amounts of equal value to rename a number
Example: 5 + 8 = 13 ones or 1 ten 3 ones

regular polygon [reg'yə•lər päl'i•gän] **polígono regular** A polygon that has all sides that are equal in length and all angles equal in measure
Examples:

related facts [ri•lāt'id fakts] **operaciones relacionadas** A set of related addition and subtraction, or multiplication and division, number sentences
Examples: 4 × 7 = 28   28 ÷ 4 = 7
           7 × 4 = 28   28 ÷ 7 = 4

remainder [ri•mān'dər] **residuo** The amount left over when a number cannot be divided equally

rhombus [räm'bəs] **rombo** A quadrilateral with two pairs of parallel sides and four sides of equal length
Example:

right angle [rīt ang'gəl] **ángulo recto** An angle that forms a square corner
Example:

right triangle [rīt trī′ang•gəl] **triángulo rectángulo** A triangle with one right angle
Example:

round [round] **redondear** To replace a number with another number that tells about how many or how much

rule [rool] **regla** A procedure (usually involving arithmetic operations) to determine an output value from an input value

scale [skāl] **escala** A series of numbers placed at fixed distances on a graph to help label the graph

second (sec) [sek′ənd] **segundo (seg)** A small unit of time
1 minute = 60 seconds

simplest form [sim′pləst fôrm] **mínima expresión** A fraction is in simplest form when the numerator and denominator have only 1 as a common factor

solid shape [sä′lid shāp] **cuerpo geométrico** See *three-dimensional figure.*

square [skwâr] **cuadrado** A quadrilateral with two pairs of parallel sides, four sides of equal length, and four right angles
Example:

square unit [skwâr yoo′nit] **unidad cuadrada** A unit of area with dimensions of 1 unit × 1 unit

standard form [stan′dərd fôrm] **forma normal** A way to write numbers by using the digits 0–9, with each digit having a place value *Example:* 3,540 ← standard form

straight angle [strāt ang′gəl] **ángulo llano** An angle whose measure is 180°
Example:

subtraction [səb•trak′shən] **resta** The process of finding how many are left when a number of items are taken away from a group of items; the process of finding the difference when two groups are compared; the opposite operation of addition

sum [sum] **suma o total** The answer to an addition problem

survey [sûr′vā] **encuesta** A method of gathering information

tally table [tal′ē tā′bəl] **tabla de conteo** A table that uses tally marks to record data

> **Word History**
>
> Some people keep score in card games by making marks on paper (IIII). These marks are known as tally marks. The word *tally* is related to *tailor*, from the Latin *talea*, meaning "twig." In early times, a method of keeping count was by cutting marks into a piece of wood or bone.

temperature [tem′pər•ə•chər] **temperatura** The degree of hotness or coldness usually measured in degrees Fahrenheit or degrees Celsius

tenth [tenth] **décimo** One of ten equal parts
Example:

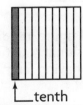

tenth

term [tûrm] **término** A number or object in a pattern

thousands [thou′zəndz] **miles** The period after the ones period in the base-ten number system

Glossary H11

three-dimensional [thrē də•men′shə•nəl] **tridimensional** Measured in three directions, such as length, width, and height
Example:

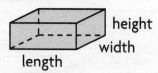

three-dimensional figure [thrē də•men′shə•nəl fig′yər] **figura tridimensional** A figure having length, width, and height

ton (T) [tun] **tonelada (t)** A customary unit used to measure weight
1 ton = 2,000 pounds

trapezoid [trap′i•zoid] **trapecio** A quadrilateral with at least one pair of parallel sides
Examples:

triangle [trī′ang•gəl] **triángulo** A polygon with three sides and three angles
Examples:

two-dimensional [tōō də•men′shə•nəl] **bidimensional** Measured in two directions, such as length and width
Example:

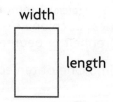

two-dimensional figure [tōō də•men′shə•nəl fig′yər] **figura bidimensional** A figure that lies in a plane; a shape having length and width

unit fraction [yōō′nit frak′shən] **fracción unitaria** A fraction that has a numerator of one

variable [vâr′ē•ə•bəl] **variable** A letter or symbol that stands for a number or numbers

Venn diagram [ven dī′ə•gram] **diagrama de Venn** A diagram that shows relationships among sets of things
Example:

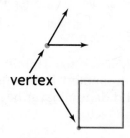

vertex [vûr′teks] **vértice** The point at which two rays of an angle meet or two (or more) line segments meet in a two-dimensional shape
Examples:

vertical [vûr′ti•kəl] **vertical** In the direction from top to bottom

weight [wāt] **peso** How heavy an object is

whole [hōl] **entero** All of the parts of a shape or group

word form [wûrd fôrm] **en palabras** A way to write numbers by using words
Example: Four hundred fifty-three thousand, two hundred twelve

yard (yd) [yärd] **yarda (yd)** A customary unit for measuring length or distance
1 yard = 3 feet

Zero Property of Multiplication [zē′rō präp′ər•tē əv mul•tə•pli•kā′shən] **propiedad del cero de la multiplicación** The property that states that the product of 0 and any number is 0
Example: $0 \times 8 = 0$

Corrections

 COMMON CORE STATE STANDARDS

Standards You Will Learn

Mathematical Practices		Some examples are:
MP1	Make sense of problems and persevere in solving them.	Lessons 1.8, 2.1, 3.2, 4.6, 5.3, 6.5, 7.4, 9.2, 10.4, 11.5, 12.8, 13.5
MP2	Reason abstractly and quantitatively.	Lessons 1.3, 2.2, 3.3, 4.8, 5.6, 6.2, 7.7, 8.2, 9.5, 10.5, 12.3, 13.4
MP3	Construct viable arguments and critique the reasoning of others.	Lessons 1.7, 2.10, 4.10, 5.2, 6.7, 7.3, 10.5, 11.1, 12.2
MP4	Model with mathematics.	Lessons 1.2, 2.5, 3.1, 4.7, 5.1, 6.4, 7.1, 8.3, 9.4, 10.1, 12.5
MP5	Use appropriate tools strategically.	Lessons 1.1, 2.3, 4.1, 4.7, 6.1, 9.1, 10.7, 11.3, 12.9
MP6	Attend to precision.	Lessons 1.4, 2.7, 3.6, 4.5, 6.3, 7.5, 8.5, 9.3, 10.2, 11.4, 12.4, 13.3
MP7	Look for and make use of structure.	Lessons 1.5, 2.4, 3.5, 4.4, 5.4, 6.6, 7.6, 8.1, 9.7, 10.3, 11.2, 12.1, 13.2
MP8	Look for and express regularity in repeated reasoning.	Lessons 1.6, 3.4, 4.3, 7.9, 8.4, 9.6, 10.6, 12.10, 13.1

Standards You Will Learn

Student Edition Lessons

Domain: Operations and Algebraic Thinking

Use the four operations with whole numbers to solve problems

4.OA.A.1	Interpret a multiplication equation as a comparison, e.g., interpret $35 = 5 \times 7$ as a statement that 35 is 5 times as many as 7 and 7 times as many as 5. Represent verbal statements of multiplicative comparisons as multiplication equations.	Lesson 2.1
4.OA.A.2	Multiply or divide to solve word problems involving multiplicative comparison, e.g., by using drawings and equations with a symbol for the unknown number to represent the problem, distinguishing multiplicative comparison from additive comparison.	Lessons 2.2, 4.12
4.OA.A.3	Solve multistep word problems posed with whole numbers and having whole-number answers using the four operations, including problems in which remainders must be interpreted. Represent these problems using equations with a letter standing for the unknown quantity. Assess the reasonableness of answers using mental computation and estimation strategies including rounding.	Lessons 2.9, 2.12, 3.7, 4.3

Gain familiarity with factors and multiples.

4.OA.B.4	Find all factor pairs for a whole number in the range 1–100. Recognize that a whole number is a multiple of each of its factors. Determine whether a given whole number in the range 1–100 is a multiple of a given one-digit number. Determine whether a given whole number in the range 1–100 is prime or composite.	Lessons 5.1, 5.2, 5.3, 5.4, 5.5

Generate and analyze patterns.

4.OA.C.5	Generate a number or shape pattern that follows a given rule. Identify apparent features of the pattern that were not explicit in the rule itself.	Lessons 5.6, 10.7

Correlations H15

Standards You Will Learn

Student Edition Lessons

Domain: Number and Operations in Base Ten

Generalize place value understanding for multi-digit whole numbers.

4.NBT.A.1	Recognize that in a multi-digit whole number, a digit in one place represents ten times what it represents in the place to its right.	Lessons 1.1, 1.5
4.NBT.A.2	Read and write multi-digit whole numbers using base-ten numerals, number names, and expanded form. Compare two multi-digit numbers based on meanings of the digits in each place, using >, =, and < symbols to record the results of comparisons.	Lessons 1.2, 1.3
4.NBT.A.3	Use place value understanding to round multi-digit whole numbers to any place.	Lesson 1.4

Use place value understanding and properties of operations to perform multi-digit arithmetic.

4.NBT.B.4	Fluently add and subtract multi-digit whole numbers using the standard algorithm.	Lessons 1.6, 1.7, 1.8
4.NBT.B.5	Multiply a whole number of up to four digits by a one-digit whole number, and multiply two two-digit numbers, using strategies based on place value and the properties of operations. Illustrate and explain the calculation by using equations, rectangular arrays, and/or area models.	Lessons 2.3, 2.4, 2.5, 2.6, 2.7, 2.8, 2.10, 2.11, 3.1, 3.2, 3.3, 3.4, 3.5, 3.6
4.NBT.B.6	Find whole-number quotients and remainders with up to four-digit dividends and one-digit divisors, using strategies based on place value, the properties of operations, and/or the relationship between multiplication and division. Illustrate and explain the calculation by using equations, rectangular arrays, and/or area models.	Lessons 4.1, 4.2, 4.4, 4.5, 4.6, 4.7, 4.8, 4.9, 4.10, 4.11

H16 Correlations

Standards You Will Learn

Student Edition Lessons

Domain: Number and Operations—Fractions

Extend understanding of fraction equivalence and ordering.

4.NF.A.1	Explain why a fraction a/b is equivalent to a fraction $(n \times a)/(n \times b)$ by using visual fraction models, with attention to how the number and size of the parts differ even though the two fractions themselves are the same size. Use this principle to recognize and generate equivalent fractions.	Lessons 6.1, 6.2, 6.3, 6.4, 6.5
4.NF.A.2	Compare two fractions with different numerators and different denominators, e.g., by creating common denominators or numerators, or by comparing to a benchmark fraction such as 1/2. Recognize that comparisons are valid only when the two fractions refer to the same whole. Record the results of comparisons with symbols >, =, or <, and justify the conclusions, e.g., by using a visual fraction model.	Lessons 6.6, 6.7, 6.8

Correlations H17

Standards You Will Learn

Student Edition Lessons

Domain: Number and Operations–Fractions

Build fractions from unit fractions by applying and extending previous understandings of operations on whole numbers.

4.NF.B.3	Understand a fraction a/b with $a > 1$ as a sum of fractions $1/b$.	
	a. Understand addition and subtraction of fractions as joining and separating parts referring to the same whole.	Lesson 7.1
	b. Decompose a fraction into a sum of fractions with the same denominator in more than one way, recording each decomposition by an equation. Justify decompositions, e.g., by using a visual fraction model.	Lessons 7.2, 7.6
	c. Add and subtract mixed numbers with like denominators, e.g., by replacing each mixed number with an equivalent fraction, and/or by using properties of operations and the relationship between addition and subtraction.	Lessons 7.7, 7.8, 7.9
	d. Solve word problems involving addition and subtraction of fractions referring to the same whole and having like denominators, e.g., by using visual fraction models and equations to represent the problem.	Lessons 7.3, 7.4, 7.5, 7.10
4.NF.B.4	Apply and extend previous understandings of multiplication to multiply a fraction by a whole number.	
	a. Understand a fraction a/b as a multiple of $1/b$.	Lesson 8.1
	b. Understand a multiple of a/b as a multiple of $1/b$, and use this understanding to multiply a fraction by a whole number.	Lessons 8.2, 8.3
	c. Solve word problems involving multiplication of a fraction by a whole number, e.g., by using visual fraction models and equations to represent the problem.	Lessons 8.4, 8.5

Standards You Will Learn

Student Edition Lessons

Domain: Number and Operations—Fractions

Understand decimal notation for fractions, and compare decimal fractions.

4.NF.C.5	Express a fraction with denominator 10 as an equivalent fraction with denominator 100, and use this technique to add two fractions with respective denominators 10 and 100.	Lessons 9.3, 9.6
4.NF.C.6	Use decimal notation for fractions with denominators 10 or 100.	Lessons 9.1, 9.2, 9.4
4.NF.C.7	Compare two decimals to hundredths by reasoning about their size. Recognize that comparisons are valid only when two decimals refer to the same whole. Record the results of comparisons with the symbols >, =, or <, and justify the conclusions, e.g., by using a visual model.	Lesson 9.7

Domain: Measurement and Data

Solve problems involving measurement and conversion of measurements from a larger unit to a smaller unit.

4.MD.A.1	Know relative sizes of measurement units within one system of units including km, m, cm; kg, g; lb, oz.; l, ml; hr, min, sec. Within a single system of measurement, express measurements in a larger unit in terms of a smaller unit. Record measurement equivalents in a two-column table.	Lessons 12.1, 12.2, 12.3, 12.4, 12.6, 12.7, 12.8, 12.11
4.MD.A.2	Use the four operations to solve word problems involving distances, intervals of time, liquid volumes, masses of objects, and money, including problems involving simple fractions or decimals, and problems that require expressing measurements given in a larger unit in terms of a smaller unit. Represent measurement quantities using diagrams such as number line diagrams that feature a measurement scale.	Lessons 9.5, 12.9, 12.10
4.MD.A.3	Apply the area and perimeter formulas for rectangles in real world and mathematical problems.	Lessons 13.1, 13.2, 13.3, 13.4, 13.5

Correlations H19

Standards You Will Learn

Student Edition Lessons

Domain: Measurement and Data

Represent and interpret data.

4.MD.B.4	Make a line plot to display a data set of measurements in fractions of a unit (1/2, 1/4, 1/8). Solve problems involving addition and subtraction of fractions by using information presented in line plots.	Lesson 12.5

Geometric measurement: understand concepts of angle and measure angles.

4.MD.C.5	Recognize angles as geometric shapes that are formed wherever two rays share a common endpoint, and understand concepts of angle measurement:	
	a. An angle is measured with reference to a circle with its center at the common endpoint of the rays, by considering the fraction of the circular arc between the points where the two rays intersect the circle. An angle that turns through 1/360 of a circle is called a "one-degree angle," and can be used to measure angles.	Lessons 11.1, 11.2
	b. An angle that turns through n one-degree angles is said to have an angle measure of n degrees.	Lesson 11.2
4.MD.C.6	Measure angles in whole-number degrees using a protractor. Sketch angles of specified measure.	Lesson 11.3
4.MD.C.7	Recognize angle measure as additive. When an angle is decomposed into non-overlapping parts, the angle measure of the whole is the sum of the angle measures of the parts. Solve addition and subtraction problems to find unknown angles on a diagram in real world and mathematical problems, e.g., by using an equation with a symbol for the unknown angle measure.	Lessons 11.4, 11.5

H20 Correlations

Standards You Will Learn

Student Edition Lessons

Domain: Geometry

Draw and identify lines and angles, and classify shapes by properties of their lines and angles.

4.G.A.1	Draw points, lines, line segments, rays, angles (right, acute, obtuse), and perpendicular and parallel lines. Identify these in two-dimensional figures.	Lessons 10.1, 10.3
4.G.A.2	Classify two-dimensional figures based on the presence or absence of parallel or perpendicular lines, or the presence or absence of angles of a specified size. Recognize right triangles as a category, and identify right triangles.	Lessons 10.2, 10.4
4.G.A.3	Recognize a line of symmetry for a two-dimensional figure as a line across the figure such that the figure can be folded along the line into matching parts. Identify line-symmetric figures and draw lines of symmetry.	Lessons 10.5, 10.6

Common Core State Standards (c) Copyright 2010. National Governors Association Center for Best Practices and Council of Chief State School Officers. All rights reserved. This product is not sponsored or endorsed by the Common Core State Standards Initiative of the National Governors Association Center for Best Practices and the Council of Chief State School Officers.

Correlations   H21

Index

A

A.M., 691–693
Activities
   Activity, 5, 279, 311, 339, 345, 359, 549, 550, 555, 556, 562, 567, 568, 575, 581, 582, 613, 614, 647, 653, 686, 703,
   Investigate, 31, 87, 157, 203, 227, 235, 247, 327, 385, 601, 621, 673
Acute angles, 550, 555
Acute triangles, 555–558
Addition
   Additive comparison, 49–52, 63
   bar models, 49–52, 70–71
   draw a diagram, 49–52
   draw a model, 70–71
   equations, 50, 385–388, 409–412
   estimating sums, 38–40
   fractions
      fractional parts of 10 and 100, 527–530
      like denominators, 391–394, 409–412, 423–426, 435–438, 527–530
      modeling, 385–388, 397–400, 409–412, 423–426, 435–438, 441–444
      and properties of addition, 435–438
      unit fractions, 391–394
   hundreds, 37–40
   hundred thousands, 37–40
   as inverse operation, 44, 50
   mixed numbers, 423–426
   modeling, 49–52, 70–71
   number patterns, 311–314
   ones, 37–40
   problem solving, 49–52
   properties
      Associative Property of Addition, 39, 435–438
      Commutative Property of Addition, 39, 435–438
   regrouping, 37–40
   tens, 37–40
   ten thousands, 37–40
   thousands, 37–40
   whole numbers, 37–40

Algebra
   addition
      Associative Property of Addition, 39, 435–438
      Commutative Property of Addition, 39, 435–438
   area
      finding, 723–726, 729–732, 743–746
      formula for, 723–726
   Distributive Property, 87–90, 99, 227–230, 718
      division and, 227–230
      multiplication and, 87–90, 99–102, 108–109
   division related to multiplication, 228, 475, 476
   inverse operations, 260
   equations
      addition, 50, 385–388, 397–400, 409–412
      angle equations, 622–623, 627–630, 631–632
      division, 210–223, 738
      multiplication, 63–66, 131–134, 470, 475–478
      multistep problems and, 131–134
      representing, 69–72, 131–134, 157–159, 203–204, 391, 394
      solving, using mental math, 107–110
      subtraction, 49–52, 132, 409–412
      writing, 49–51, 63–66, 69–72, 386–387, 391–394
   expressions
      function tables and, 286–287
      numerical, 435–438
      with parentheses, 107, 145
   measurement units, converting, 647–650, 653–656, 659–662, 673–676, 679–682, 685–688
   multiplication
      Associative Property of Multiplication, 107–110, 145–148
      Commutative Property of Multiplication, 63–66, 107–110, 171
      comparison problems, 63–66, 69–72
      Distributive Property, 87–90, 99–101, 108–109, 227–230, 718
      finding unknown factors, 285–288
      related to division, 475

number patterns, 311–314, 470–472
patterns in measurement units, 703–706
perimeter
  finding, 717–720
  formula for, 717–720
writing a rule, 121

Algebraic thinking, operations and, 63, 69, 131–134, 265–268

Algorithms
  addition, 37–40
  division, 253–256, 259–262
  multiplication, 119–122, 125–128, 163–166, 171–174
  subtraction, 43–46

Analog clocks, 602, 685–688

Angles
  acute, 550, 555, 608, 615
  angle equations, 622–623, 627–630, 631–632
  defined, 550
  classify and draw, 549–552
  measure and draw, 613–616
  obtuse, 550, 555, 608, 615
  right, 550, 555, 567, 608, 723
  straight, 550, 608
  turns in a circle, 607–610
  using to classify triangles, 555–558

Area
  of combined rectangles, 729–732
  concept of, 723–726
  defined, 723
  finding, 723–726, 729–732, 743–746
  find unknown measures, 737–740
  formula for, 723–726
  measure to find, 729–732
  perimeter related to, 731
  of a rectangle, 723–726, 729–732
  square unit, 723, 729
  units of, 723–726

Area models
  decimals and, 495–498, 501–504, 507, 513–514
  fractions and, 327–330, 333–336, 351–354, 365
  multiplication and, 157–160, 163–164

Arrays, 279–282

Art, Connect to, 400, 570

Assessment
  Diagnostic Assessment
    Show What You Know, 3, 61, 143, 195, 277, 325, 383, 453, 493, 547, 599, 639, 715

  Ongoing Assessment, Mid-Chapter Checkpoint, 29–30, 105–106, 169–170, 233–234, 297–298, 357–358, 415–416, 467–468, 525–526, 573–574, 619–620, 671–672, 735–736
  Summative Assessment
    Chapter Review/Test, 55–60, 137–142, 189–194, 271–276, 317–322, 377–382, 447–452, 487–492, 539–544, 593–598, 633–638, 709–714, 749–754

Associative Property
  of Addition, 39, 435–438
  of Multiplication, 107–109, 145–147

Bar graphs, 116
Bar models, 49–52, 183–186, 265–268, 481–484, 627–630
Base, 723–726
Base-ten blocks, 5, 31, 247–249
Benchmarks, 26. See also Measurement and Data
  defined, 360
  fraction, 359–362
  measurement, 641–644
Break-apart strategy, 87, 99, 157, 163, 227

Centimeters, 673–676, 717–720
Cents, 513–516
Chapter Openers, 3, 61, 143, 195, 277, 325, 383, 453, 493, 547, 599, 639, 715
Chapter Review/Test, 55–60, 137–142, 189–194, 271–276, 317–322, 377–382, 447–452, 487–492, 539–544, 593–598, 633–638, 709–714, 749–754
Checkpoint, Mid-Chapter. See Mid-Chapter Checkpoint
Classifying
  angles, 549–552
  lines, 549–552
  quadrilaterals, 567–570
  triangles, 555–558

Index **H23**

Clocks
　analog, 602, 685–688
　elapsed time, 691–694
Clockwise, 601–604
Combined rectangles, 729–732
Common Core State Standards, H14–H21
Common denominators, 345–348, 365–368, 409–412, 527–530
Common factors, 291–294
　fractions in simplest form using, 340
Common multiples, 299–302, 345–348
Common numerators, 365–368
Communicate Math Ideas
　Math Talk, In every Student Edition lesson. Some examples are: 5, 12, 64, 82, 114, 132, 177, 209, 228, 279, 299, 328, 359, 386, 429, 455, 495, 520, 556, 602, 642, 723
　Read Math, 555, 685
　Write Math, In every Student Edition lesson. Some examples are: 8, 40, 96, 116, 224, 288, 348, 420, 510, 530, 682
Commutative Property
　of Addition, 39, 435–438
　of Multiplication, 63–66, 107–110, 171
Comparing
　decimals, 533–536
　fractions, 359–362, 365–368, 371–374
　measurement units, 647–650, 653–656, 659–662, 673–676, 679–682, 685–688
　whole numbers, 17–20
Comparison problems, 49–52, 69–72, 481–484
Compatible numbers, 152, 221–223
Composite numbers, 305–307
Connect, 99, 163, 260, 435, 607, 703
Connect to Art, 400, 424
Connect to Reading, 84, 224
Connect to Science, 26, 218, 510, 616, 688, 740
Connect to Social Studies, 308
Correlations
　Common Core State Standards, H14–H21
Counterclockwise, 601–604
Counters, 203–206, 235, 236
Counting numbers, 197, 455–458, 461–464
Critical Area, 1, 323, 545

Cross-Curricular Activities and Connections
　Connect to Art, 400, 570
　Connect to Reading, 84, 224
　Connect to Science, 26, 218, 510, 616, 688, 740
　Connect to Social Studies, 308
Cup, 659
Customary units
　benchmarks, 641-644
　converting, 647–650, 653–656, 659–662
　of distance, 641
　of length, 641, 647–650
　of liquid volume, 641, 659–662
　of weight, 641, 653–656

Data
　gathering, 26
　using
　　bar graphs, 116
　　line plots, 665–668
　　tables. *See* Tables
　　tally tables, 665–668, 671
　　Venn diagrams, 196, 556, 558, 568, 716
Days, 686
Decimal point, 495
Decimals
　comparing
　　using models, 533–536
　　using place value, 533–536
　defined, 495
　equivalent decimals
　　defined, 508
　　modeling, 507–510
　hundredths, 501–504, 507–510
　place value and, 495–498, 501–504, 534–536
　relating
　　to fractions, 495–498, 501–504, 507–510, 513–516, 527–530
　　to mixed numbers, 495–497, 501–503, 509, 514–515
　　to money, 513–516
　tenths, 495–498, 507–510
Decimeters, 673
Degrees
　angle measures and, 607–610
Denominators, 365–368

H24　Index

Diagrams. *See also* Graphic Organizers
  Draw a Diagram, 49–52, 72, 113–116, 183–186, 265–268, 481–484, 627–630, 691–694
  using, 49–52, 72, 113–116, 158, 183–186, 265–268, 481–484, 627–630, 691–694, 729–732
  Venn diagrams, 196, 556, 558, 568, 716

Difference, 43–44, 195, 403–405, 410–411, 423–425, 430–431

Differentiated Instruction
  Show What You Know, 3, 61, 143, 195, 277, 325, 383, 453, 493, 547, 599, 639, 715

Digit, 6
  value of, 5–8

Distance, 641–642

Distributive Property, 87–90, 99–100, 227–230, 718
  division using, 227–230
  models, 87–90, 99–101
  multiplication using, 87–90, 99–101

Dividend, 209. *See also* Division

Divisibility, 285–288, 306

Division
  basic facts, 215–216, 221–222, 259–262
  Distributive Property, 227–230
  dividend, 209
  divisor, 209
  equations, 210–212, 215–218, 222–223
  estimating quotients
    using compatible numbers, 221–224
    using multiples, 197–200
  to find equivalent fractions, 340–341
  inverse relationship to multiplication, 260
  modeling
    with bar models, 265–268
    with base-ten blocks, 247–249
    with counters, 203–205, 235
    draw a diagram, 265–268
    with fraction circles, 404–406
    with fraction strips, 404–405, 409
    with quick pictures, 203–206, 209–212
  of multidigit numbers, 253–256, 259–262
  partial quotients, 241–244
  patterns, 216–217, 285–287
  placing the first digit of the quotient, 253–256
  quotient, 209
  regrouping, 247–250, 253–256, 259–262
  remainders, 203–206
    defined, 204
    interpreting, 209–212
  tens, hundreds, thousands, 215–218
  types of problems, 197–200
  using repeated subtraction, 235–238
  with zero, 215
  zero in the quotient, 260

Divisor, 209. *See also* Division

Dollar, 513–514

Draw a Diagram. *See* Problem-solving strategies

Draw Conclusions, 31–32, 88, 158, 204, 228, 236, 247, 328, 386, 602, 622, 673–674

E

Elapsed time
  draw a diagram, 691–694
  find, 691–694

Equal sign, 17–18, 328

Equations
  addition, 50, 385–388, 409–412
  angle equations, 623, 630, 633
  division, 209–223, 738
  inverse operations and, 260
  multiplication, 63–66, 131–134, 470, 475–478
  multistep problems and, 131–134
  representing, 69–72, 131–134, 157–159, 203–204, 391, 394
    using mental math, 107–110
  subtraction, 49–52, 132, 409–412
  writing, 49–51, 63–66, 69–72, 386–387, 391–394

Equivalent decimals
  defined, 508
  model, 507–510

Equivalent fractions
  defined, 327
  modeling, 327–330, 333–336, 339–342, 351–354
  multiplication and division to find, 333–336, 340

Error Alert, 38, 260, 305, 346, 582, 614, 698, 738

Errors
  Error Alert, 38, 260, 305, 346, 582, 614, 698, 738
  What's the Error?, 13, 46, 96, 134, 206, 330, 409, 706

Essential Question, In every Student Edition lesson. Some examples are: 5, 31, 69, 107, 145, 197, 227, 279, 327, 385, 455, 495, 601, 665, 737

Estimate. *See also* Benchmarks; Rounding
  defined, 23
  products, 81–83, 151–154
  quotients
    using compatible numbers, 221–223
    using multiples, 197–200
  reasonableness of
    addition, 38–39
    fractions, 371
    multiplication, 151–154
    subtraction, 43
  rounding to find an, 23–26
  using place value, 24

Expanded form
  to multiply, 93–96
  whole numbers, 11–14

Expressions
  function tables and, 285–287
  numerical, 435–438
  with parentheses, 107, 145

F

Factors
  common, 291–294
  defined, 279
  and divisibility, 285–288
  find unknown, 737
  list all factor pairs, 286, 287
  model, 279–282
  multiples and, 299–302

Feet, 647–650

Find a Pattern. *See* Problem-solving strategies

Fluid ounces, 659–660

Formulas
  for area, 723–726
  for perimeter, 717–720

Four-digit numbers
  multiplication, by one-digit, 125–128

Fractions
  addition
    like denominators, 391–394, 409–412, 423–426, 435–438, 527–530
    mixed numbers, 423–426
    modeling, 385–388, 397–400, 409–412, 423–426, 441–444
    parts of 10 and 100, 527–530
    properties, 435–438
    of unit fractions, 391–394
  area models, 327–330, 333–336, 351–354
  bar models, 481–484
  benchmarks, 359–362
  common denominator, 345–348, 365–368, 410
  common numerator, 365–368
  comparing
    using benchmarks, 359–362, 371–374
    using common denominators, 365–368, 371–374
    using common numerators, 365–368, 371–374
    using fraction strips, 359–361
    using models, 365
    using number lines, 371–374
    using simplest form, 366–367
  comparison problems, 481–484
  denominator, 365–368
  division and, 476
  draw a diagram, 481–484
  equivalent fractions
    and common denominators, 346–347
    defined, 327
    modeling, 327–330, 333–336, 339–342, 351–354
    multiplication and division to find, 333–336
  greater than 1, 417–420, 429–432
  mixed numbers and, 417–420
  multiples of, 455–458, 461–464
  multiplication with whole numbers, 469–472
    using addition, 469–472
    using models, 469–472
    using patterns, 470–471
  multistep problems, 441–444
  numerator, 365–368
  ordering, 371–374
  patterns, 470
  relating
    to decimals, 495–498, 501–504, 508, 513–516, 527–530
    to money, 513–516

H26 Index

renaming as mixed numbers, 417–420, 476
simplest form, 339–342
subtraction
  like denominators, 409–412, 423–426
  mixed numbers, 423–426
  modeling, 385–388, 403–406, 409–412, 429–432
  with renaming, 429–432

Gallon, 659–662
Geometry. *See also* Two-dimensional figures
  angles, 549–552
    acute, 550, 555, 608, 615
    angle equations, 622–623, 627–630, 631–632
    classify triangles by size of, 555–558
    find unknown measures, 737–740
    measure and draw, 613–616
    obtuse, 550, 555, 608, 615
    right, 550, 555, 567, 608, 723
    turns on a clock and angle measures, 602, 603, 685
  hexagons, 581
  lines, 549–552
    parallel, 561–564
    perpendicular, 561–564
  polygons, 555–558, 567–570
  quadrilaterals, 567–570
    classifying, 567–570
  rays, 549–552
  rectangles, 567–569
  shape patterns, 587–590
  squares, 567–569, 581
  symmetry
    lines of symmetry, 581–584
    line symmetry, 575–578, 581–584
  trapezoids, 567–569, 581
  triangles, 555–558
    classifying, 555–558
Glossary, Multimedia. *See* Multimedia eGlossary
Go Deeper, In some Student Edition lessons. Some examples are: 19, 90, 229, 262, 294, 388, 444, 463, 536, 609, 687
Grams, 679–682
Graphic Organizers. *See also* Tables
  Brainstorming Diagram, 640
  Bubble Map, 384, 454, 600
  Flow Map, 62, 278, 326, 548
  H-Diagram, 144
  problem solving, 49–51, 113–114, 183–184, 291–292, 351–352, 441–442, 481–482, 519–520, 587–588, 627–628, 691–692, 743–744
  Semantic Map, 494
  Venn diagram, 196, 556, 558, 568, 716
Graphs. *See also* Data
  bar graphs, 116
  line plots, 665–668
  Venn diagrams, 196, 556, 558, 568, 716
Greater than (>), 17–18, 151, 153, 259
Grid, 38, 44, 87, 89, 157, 227, 229, 235, 305, 306, 729

Half gallon, 659
Half hour, 685–688
Height, 723–726, 737, 739
Hexagons, 581
Hours
  elapsed time and, 691–694
  half, 685–688
  minutes after the hour, 685, 692
Hundreds, 11–14
  addition, 37–40
  multiply by whole numbers, 75–78, 125
  place value, 5–8, 11–14, 37–40
  subtraction, 43–46
Hundred thousands, 11–14
  addition, 37–40
  place value, 5–8, 11–14, 37–40
  rounding, 23
  subtraction, 43–46
Hundredths, 501–504, 507–510

Identity Property of Multiplication, 476
Inches, 647–650
Interpret remainders, 209–212

Intersecting lines, 561–564
Intervention
    Show What You Know, 3, 61, 143, 195, 277, 325, 383, 453, 493, 547, 599, 639, 715
Introduce the Chapter, 3, 61, 143, 195, 277, 325, 383, 453, 493, 547, 599, 639, 715
Inverse operations
    addition and subtraction, 44, 50
    multiplication and division, 260
Investigate, 31, 87–88, 157–158, 203–204, 227–228, 235–236, 247–248, 327–328, 385–386, 601–602, 621–622
iTools. *See* Technology and Digital Resources

Kilograms, 679–682
Kilometers, 642, 643

Length
    area and unknown measures, 737–740
    convert customary units, 641, 647–650
    convert metric units, 673–676
    customary units for, 641, 647–650
    measuring and estimating, 641–644, 673–676
    metric units for, 642, 673–676
    perimeter and unknown measures, 737–740
Less than (<), 17–18, 151, 153, 253, 469
Line plots, 665–668
Line segment, 549–552, 723
Lines of symmetry, 581–584
Line symmetry, 575–578, 581–584
Linear models. *See* Number lines
Lines, 549–552
    draw, 562–563
    intersecting, 561–564
    parallel, 561–564
    perpendicular, 561–564, 723

Liquid volume, 641–642, 659–662, 679–682, 704
Liters, 679–682

Make a Table or List. *See* Problem-solving strategies
Make Connections, 32, 88, 158, 204, 228, 236, 248, 328, 386, 602, 622, 674
Manipulatives and materials
    base-ten blocks, 5–8, 31, 247–250
    bills and coins, 519–522
    color pencils, 87, 227, 327, 339, 385, 555
    construction paper, 621
    counters, 203–206, 235, 236
    decimal models, 513–514
    dot paper, 581
    fraction circles, 385, 601
    fraction strips, 359, 397–398, 404
    grid paper, 87, 163, 227, 235, 647, 729
    MathBoard. *See* MathBoard
    number line, 18, 23, 25
    pattern blocks, 575–576, 582
    place-value chart, 6, 11, 17, 32–33, 495–496, 501–502
    protractors, 613–616, 621–627
    rulers, centimeter, 673
    scissors, 575, 621
    square tiles, 279
    tracing paper, 575
Mass
    converting metric units, 679–682
    measuring, 679–682
    metric units for, 642, 679–682
MathBoard, In every Student Edition lesson. Some examples are: 7, 44, 100, 132, 172, 198, 222, 280, 329, 387, 457, 497, 551, 603, 643, 719
Mathematical Practices
    1. Make sense of problems and persevere in solving them. In many lessons. Some examples are: 43, 49, 63, 81, 113, 125, 131, 151, 163, 171, 183, 197, 227, 247, 265, 291, 351, 403, 423, 429, 441, 475, 481, 501, 567, 581, 601, 613, 621, 627, 641, 673, 685, 691, 717, 729, 737, 743

2. Reason abstractly and quantitatively. In many lessons. Some examples are: 11, 17, 23, 31, 63, 69, 75, 113, 119, 131, 145, 151, 157, 163, 221, 241, 247, 279, 297, 311, 333, 359, 365, 385, 391, 397, 403, 409, 417, 423, 435, 441, 461, 495, 501, 507, 513, 519, 527, 533, 567, 575, 601, 607, 641, 647, 653, 665, 723, 737

3. Construct viable arguments and critique the reasoning of others. In many lessons. Some examples are: 43, 87, 119, 125, 131, 209, 221, 253, 285, 365, 391, 397, 441, 575, 581, 601, 641, 647, 659, 665, 679, 697, 703

4. Model with mathematics. In many lessons. Some examples are: 11, 23, 49, 87, 99, 113, 145, 157, 183, 227, 247, 265, 279, 285, 327, 333, 345, 351, 385, 403, 429, 441, 455, 469, 513, 549, 561, 621, 627, 647, 659, 665, 673, 685, 691, 743

5. Use appropriate tools strategically. In many lessons. Some examples are: 5, 31, 75, 197, 215, 235, 259, 265, 311, 327, 333, 365, 495, 587, 613, 685, 691

6. Attend to precision. In many lessons. Some examples are: 5, 23, 37, 63, 87, 99, 125, 151, 157, 177, 215, 221, 235, 241, 299, 339, 345, 351, 359, 403, 409, 417, 455, 495, 507, 513, 519, 533, 555, 561, 587, 607, 613, 653, 659, 703, 723, 729

7. Look for and make use of structure. In many lessons. Some examples are: 11, 31, 75, 81, 119, 145, 171, 177, 197, 209, 215, 227, 253, 285, 299, 311, 327, 339, 345, 359, 409, 417, 429, 455, 461, 481, 501, 507, 527, 555, 561, 587, 607, 641, 659, 673, 685, 703, 717, 737

8. Look for and express regularity in repeated reasoning. In many lessons. Some examples are: 37, 43, 49, 163, 171, 177, 209, 241, 259, 299, 391, 417, 423, 435, 461, 475, 527, 673, 697, 717

Math Idea, 24, 44, 253, 279, 285, 306, 371, 575, 607, 621, 673, 724

Math in the Real World Activities, 3, 61, 143, 195, 277, 325, 383, 453, 493, 547, 599, 639, 715

Math on the Spot Videos, In every Student Edition lesson. Some examples are: 8, 46, 90, 116, 148, 217, 244, 281, 330, 388, 426, 458, 498, 552, 604, 649, 720

Math Talk, In every Student Edition lesson. Some examples are: 5, 12, 64, 82, 114, 132, 177, 209, 228, 279, 299, 328, 359, 386, 429, 455, 495, 520, 556, 602, 642, 723

Measurement
  angles
    and fractional parts of circle, 601–604
    joining and separating, 621–624
    measuring and drawing, 613–616
    solving for unknown angle measures, 627–630
  area, 723–726, 743–746
    of combined rectangles, 729–732
    concept of, 723–726
    defined, 723
    finding, 723–726, 729–732, 743–746
    finding unknown measures, 737–740
    formula for, 723–726
    perimeter, related to, 731
    of a rectangle, 723–726, 729–732
    square unit, 723
    units of, 723–726
  benchmarks, 641–644
    centimeter, 642–644
    cup, 641
    decimeter, 642
    fluid ounce, 641
    foot, 641, 643
    gallon, 641, 643
    gram, 642–643
    inch, 641, 643
    kilogram, 642–643
    kilometer, 642–643
    liter, 642–643
    meter, 642–643
    mile, 641, 643
    milliliter, 642–644
    millimeter, 642
    ounce, 641, 643
    pint, 641
    pound, 641, 643
    quart, 641
    ton, 641, 643
    yard, 641, 643
  comparing, 642, 647–650, 653–656, 659–662, 685–688

concept of, 641
conversions, 641–644, 647–650, 653–656, 659–662, 673–676, 679–682, 685–688, 691–694, 697–700, 703–706, 717–720, 723–726, 729–732, 737–740, 743–746
customary units
  benchmarks, 641–644
  of distance, 641
  of length, 641, 647–650
  of liquid volumes, 641, 659–662
  of weight, 641, 653–656
degrees (of a circle), 607–610
line plots, 665–668
mass, 642, 679–682
metric units
  benchmarks, 642–644
  converting, 673–676, 679–682
  of distance, 642
  of length, 642, 673–676
  of liquid volume, 679–682
  of mass, 642, 679–682
mixed measures, 697–700
money problems, 519–522
patterns in measurement units, 703–706
perimeter, 717–720
  defined, 717
  finding, 717–720
  finding unknown measures, 737–740
  formulas for, 717–720
  measuring, 737–740
  of a rectangle, 717–720
  of a square, 717–720
time
  elapsed, 691–694
  units of, 685–688

Mental Math, 81, 107–110, 151–152, 435–438

Meters, 673–676

Metric units
benchmarks, 642–644
converting, 673–676, 679–682
of distance, 642
of length, 642, 673–676
of liquid volume, 679–682
of mass, 642, 679–682

Mid-Chapter Checkpoint, 29–30, 105–106, 169–170, 233–234, 297–298, 357–358, 415–416, 467–468, 525–526, 573–574, 619–620, 671–672, 735–736

Miles, 641–643

Milliliters, 642–644, 679–682

Millimeters, 673–676

Millions
place value and, 5–8

Minutes. See also Time
elapsed time and, 691–694

Mixed measures, 697–700

Mixed numbers
addition, 423–426
decimals, related to, 496–498, 501–503, 509, 514–515
defined, 417
modeling, 417–418, 429, 502
multiply by a whole number, 475–478
renaming as fractions, 417–420, 476
subtraction, 423–426, 429–432
subtraction with renaming, 429–432

Modeling
area models
  decimals and, 495, 513–516
  fractions and, 327–330, 333–336, 351–354, 359–362
bar models, 49–52, 183–186, 265–268, 481–484, 627–630, 659–662
Distributive Property, 87–90, 99–101, 108–109
division
  with base-ten blocks, 247–249
  with counters, 203–206, 235
  draw a diagram, 265–268
  with quick pictures, 203–206, 209–212, 247–250
  using inverse operations, 227–230, 259–262
equations, 131–134
fractions
  addition, 385–388, 391–394, 397–400, 409–412, 423–426, 435–438
  equivalent, 327–330, 333–336, 339–342, 345–348, 351–354, 507–510
  mixed numbers, 417–420, 423–426, 429–432, 435–438, 475–478
  subtraction, 403–406, 409–411, 429–432
multiples of ten, hundred, and thousand, 75–78
multiplication, 63–66, 69–72
  facts, 61, 143, 277
  by one-digit numbers, 87–90, 93–96, 99–102, 119–122
  with quick pictures, 75–78
  by two-digit numbers, 87–90, 119–122, 157–160, 163–166, 171–174, 177–180

quick pictures
    to model division, 203–206, 248–250
    to model multiplication, 75–78

Money
  relating
    to decimals, 513–516
    to fractions, 513–516

Multimedia eGlossary, 4, 62, 144, 196, 278, 326, 384, 454, 494, 548, 600, 640, 716

Multiples
  common, 299–302
  defined, 197
  estimate quotients using, 197–200
  factors and, 299–302
  of fractions, 461–464
  of unit fractions, 455–458

Multiplication
  area model, 157–160
  arrays, 61
  bar models, 183–186
  comparison problems, 63–66
  draw a diagram, 183–186, 265–268
  equations, 63–66, 131–134, 470, 475–478
  estimating products, 81–83, 99–102, 151–154
  expanded form, 93–96
  expressions, 131–134
  factors, 279–282
  to find equivalent fractions, 333–336
  four-digit numbers, 126–128
  fractions and whole numbers, 461–464, 475–478
  halving-and-doubling strategy, 108, 146
  as inverse of division, 737–740
  mental math, 107–110
  modeling, 186, 203–206
    one-digit numbers, 87–90, 93–96, 99–102, 119–122
    with quick pictures, 75–78
    two-digit numbers, 87–90, 119–122
    by two-digit numbers, 157–160, 163–166, 171–174, 177–180
  multiples of unit fractions, 455–458
  by multiples of 10, 100, and 1,000, 75–78
  by one-digit numbers, 87–90, 93–96, 99–102, 119–122
  partial products, 99–102, 157–160, 163–166, 177
  products, 76–77, 81–83, 99–102, 151–154, 157–160
  properties
    Associative Property of Multiplication, 107–110, 145–148
    Commutative Property of Multiplication, 63, 107–110, 171
    Distributive Property, 87–90, 99–101, 108–109, 227–230, 718
    Identity Property of Multiplication, 476
  quotient estimation, 197–200
  reasonableness of answers, 82–83
  regrouping in, 119–122, 125–128, 171–174, 178–179
  by tens, 145–148
  three-digit numbers, 125–128
  two-digit numbers, 87–90, 119–122
  by two-digit numbers, 157–160, 163–166, 171–174, 177–180

Multistep word problems, 113–116, 131–134, 183–186, 265–268

N

Not equal to sign (≠), 328

Number lines
  to compare decimals, 533
  to compare fractions, 371–373
  to compare numbers, 18
  to divide, 236–237
  to find multiples of fractions, 462
  to multiply, 76, 146
  to order fractions, 371–374, 665–668
  to order numbers, 18
  to round numbers, 23
  to show equivalent decimals, 501
  to show equivalent fractions and decimals, 501

Number patterns, 311–314, 587–590

Numbers. *See also* Counting numbers; Decimals; Fractions; Mixed numbers; Whole numbers
  benchmark, 359–362, 641–644
  comparing
    decimals, 533–536
    whole numbers, 17–20
  compatible, 152, 221–223
  composite, 305–307
  expanded form, whole numbers, 11–14
  factors of, 279–282, 291–294
  multiples of, 197–200, 299–302
  ordering
    fractions, 371–374
    whole numbers, 17–20
  place value, 5–8, 11–14, 17–20, 23–26
  prime, 305–307

read and write, 11–14
renaming, 31–34
rounding, 23–26
standard form, whole numbers, 11–14
word form, whole numbers, 11–14

Numbers and Operations
adding, 37–40. *See also* Addition
comparing, 17–20
  decimals, 533–536
division. *See also* Division
  and Distributive Property, 227–230
  estimating quotients using compatible numbers, 221–224
  estimating quotients using multiples, 197–200
  placing first digit, 253–256
  with regrouping, 247–250, 253–256, 259–262
  remainders, 203–206, 209–212
  tens, hundreds, and thousands, 215–218
  using bar models to solve multistep problems, 265–268
  using partial quotients, 241–244
  using repeated subtraction, 235–238
fractions. *See also* Fractions
  adding fractional parts of 10 and 100, 527–530
  common denominators, 345–348
  comparing, 359–362, 365–368, 371–374
    using benchmarks, 359–362
  comparison problems with, 481–484
  equivalent, 327–330, 333–336, 351–354
    and decimals, 507–510
  multiples of, 461–464
    unit fractions, 455–458
  multiplying by whole number, 469–472, 475–478
    using models, 469–472
  ordering, 371–374
  relating decimals, money and, 513–516
  relating hundredths and decimals, 501–504
  relating tenths and decimals, 495–498
  simplest form, 339–342
multiplication. *See also* Multiplication
  area models and partial products, 157–160
  choosing method for, 177–180
  estimating products, strategies for, 151–154

by tens, strategies for, 145–148
using Distributive Property, 87–90
using expanded form, 93–96
using mental math, 107–110
using partial products, 99–102, 163–166
using regrouping, 119–128, 171–174
ordering, 17–20
place value, 5–26, 31–52, 75–78. *See also* Place value
renaming, 31–34
rounding, 23–26, 81–84
subtracting, 43–46. *See also* Subtraction

Obtuse angles, 550–552, 555–558
Obtuse triangles, 555–558
Odd numbers, 312
One-digit numbers
multiplication
  four-digit by one-digit, 94–96, 125–128
  three-digit by one-digit, 93–96, 99–101, 125–128
  two-digit by one-digit, 87–90, 119–122
Ones
addition, 37–40
place value, 5–8
subtraction, 43–46
On Your Own, In every Student Edition lesson. Some examples are: 7, 44, 83, 121, 153, 222, 281, 335, 431, 457, 497, 551, 609, 667, 719, 746
Operations and Algebraic Thinking, 63–66, 69–72, 113–116, 131–134
division
  interpreting remainders, 209–212
  multistep problems, 265–268
factors
  common, solving problems with, 291–294
  and divisibility, 285–288
  modeling, 279–282
  and multiples, 299–302
multiplication
  comparison problems, 63–66, 69–72
  multistep problems, 113–116
  two-digit numbers, 183–186
number patterns, 311–314
prime and composite numbers, 305–308

solving multistep problems using equations, 131–134
Ordering
fractions, 371–374
whole numbers, 17–20
Order of Operations, 132
Ounces, 641, 643, 653–656

P

P.M., 691, 693
Parallel lines, 561–564
Parallelograms, 567–570
Parentheses, 107, 145
Partial products, 93–96, 99–102
defined, 88
four-digit by one-digit, 100–102
three-digit by one-digit, 99–102
two-digit by two-digit, 157–160, 163–166
Partial quotients, 241–244
Patterns
factor patterns, 285–287
finding, 311
in measurement units, 703–706
multiplication, 76–77, 470
number, 311–314
shape, 587–590
Perimeter
area, related to, 723–726
defined, 717
finding, 717–720
find unknown measures, 737–740
formulas for, 717–720
measuring, 737–740
of a rectangle, 717–720
of a square, 717–720
Period
place value, 11
Perpendicular lines, 561–564
Personal Math Trainer, In some Student Edition lessons. Some examples are: 277, 294, 314, 325, 383
Pint, 641, 659–662
Place value
addition and, 37–40
chart, 6, 11, 17, 32–33, 495–497, 501–502, 508
to compare and order numbers, 17–20

decimals and, 495–498, 534–536
estimate using, 24
hundreds, 5–8, 11–14
hundred thousands, 5–8, 11–14
millions, 5–8
number line and, 18
ones, 5–8, 11–14
period and, 11
relationships, 5–8
renaming, 31–34
rounding and, 23–26
tens, 5–8, 11–14, 100, 221
ten thousands, 5–8, 11–14
thousands, 5–8, 11–14, 75–78
Polygons, 555–558, 567–570
regular, 581
Pose a Problem, 90, 212, 230, 302, 314, 438
Pounds, 641, 653–656
Practice and Homework
In every Student Edition lesson. Some examples are: 9–10, 79–80, 231–232, 337–338, 407–408, 523–524, 605–606, 645–646, 727–728
Predictions, making, 84
Prerequisite Skills
Show What You Know, 3, 61, 143, 195, 277, 325, 383, 453, 493, 547, 599, 639, 715
Prime numbers, 305–307
Problem solving
addition and subtraction, 49–52
Problem-Solving Applications
Mental Math, 81, 107–110, 146, 151–152, 435–438
Pose a Problem, 90, 230, 302, 314, 438
Real-World Problem Solving, 8, 13, 20, 25, 40, 45, 72, 89, 90, 96, 102, 110, 122, 128, 133, 148, 154, 159, 166, 200, 205, 212, 217, 223, 229, 230, 237, 244, 249, 262, 281, 288, 302, 307, 314, 336, 342, 348, 362, 387, 411, 420, 426, 432, 457, 478, 498, 504, 510, 516, 530, 564, 570, 603, 616, 629, 644, 649, 655, 661, 675, 681, 687, 699, 706
Real-World Unlock the Problem, 11, 14, 17, 23–24, 34, 37, 43, 49–50, 63, 66, 69–70, 75–76, 78, 81, 93–94, 99–100, 107, 113–114, 119–120, 125, 131–132, 145–146, 151–152, 163–164, 171, 174, 177–178, 180, 183–184, 197–198, 209–210, 215, 221, 238, 241, 253, 256,

Index H33

259–260, 265–266, 279, 282, 285, 291–292, 299, 305, 311, 333–334, 339–340, 345, 351–352, 359–360, 365, 368, 371, 374, 391–392, 394, 397–398, 403–404, 406, 409, 417–418, 423–424, 429–430, 435–436, 455–456, 461–462, 464, 469, 472, 475–476, 481–482, 495–496, 501–502, 507–508, 513, 519–520, 527–528, 533, 536, 549–550, 555–556, 561–562, 567–568, 575–576, 578, 581–582, 587–588, 607–608, 610, 613, 626, 627, 641–642, 647–648, 653–654, 659–660, 665–666, 668, 679, 682, 685–686, 691–692, 697–698, 700, 703, 717–718, 720, 723–724, 726, 729–730, 732, 737–738, 743–744

Reason Abstractly, 387

Sense or Nonsense?, 160, 250, 388, 411, 412, 458, 476, 604, 699

Think Smarter Problems, In every Student Edition lesson. Some examples are: 8, 25, 63, 88, 122, 154, 211, 228, 262, 281, 328, 386, 426, 457, 498, 552, 610, 644, 720

Try This!, 11, 24, 44, 64, 107, 120, 146, 152, 178, 209, 210, 286, 312, 366, 372, 436, 476, 496, 514, 527, 528, 534, 555, 561, 568, 608, 654, 698, 704, 718, 724, 738

What's the Error?, 13, 46, 96, 134, 330, 409, 706

What's the Question?, 20, 342, 362, 478

Problem-solving strategies
Act It Out, 441–444, 519–522, 587–590
Draw a Diagram, 49–52, 113–116, 183–186, 265–268, 481–484, 627–630, 691–694
Make a List, 291–294
Make a Table, 351–354
Solve a Simpler Problem, 743–746

Products. *See also* Multiplication
estimating, 81–83, 99–102, 151–154
partial, 88, 99–101, 157–160, 163–166

Project, 2, 324, 546

Properties
Associative Property
of Addition, 39, 435–438
of Multiplication, 107–110, 145–148
Commutative Property
of Addition, 39, 435–438
of Multiplication, 63, 107–110, 171

Distributive Property, 87–90, 99–101, 108–109, 227–230, 718
Identity Property of Multiplication, 476

Quadrilaterals, 567–570
defined, 567
parallelogram, 567–570
rectangle, 567–570
rhombus, 567–570
square, 567–570
trapezoid, 567–570

Quart, 641, 659–662

Quarters, 513–516, 519–520

Quick pictures
to model division, 203–206, 248–250
to model multiplication, 75–78

Quotients, 210, 235–238. *See also* Division
estimating, 197–200, 221–224
partial, 241–244
placing the first digit, 253–256

Rays, 549–552

Reading
Connect to Reading, 84, 224
Read Math, 555, 685
Read/Solve the Problem, 49–50, 113–114, 183–184, 265–266, 291–292, 351–352, 441–442, 481–482, 519–520, 587–588, 627–628, 691–692, 743–744
Visualize It, 4, 62, 144, 196, 278, 326, 384, 454, 494, 548, 600, 640, 716

Real World
Problem Solving, In every Student Edition lesson. Some examples are: 8, 45, 96, 148, 200, 237, 302, 348, 411, 437, 457, 498, 552, 603, 649, 699
Unlock the Problem, In every Student Edition lesson. Some examples are: 11, 49–50, 93–94, 151–152, 209–210, 279, 345, 394, 429–430, 435–436, 441–442, 455–456, 495–496, 533, 549–550, 578, 647–648, 717–718

Reasonableness of an answer, 23, 43, 82, 93, 99, 119, 151–154, 163, 171, 221, 641

Rectangles
  area of, 723–726, 729–732
  identifying, 567–570
  perimeter of, 717–720
Rectangular model. *See* Area models
Regrouping
  addition and, 37–40
  division and, 247–250, 253–256
  multiplication and, 119–122, 125–128, 171–174, 178–179
  subtraction and, 43–46
Regular polygon, 581
Remainders, 203–206. *See also* Division
  defined, 204
  interpret, 209–212
Remember, 39, 63, 107, 145, 158, 209, 455, 476, 513, 528, 576, 608, 723
Rename
  fractions, 417–420, 476
  mixed numbers, 417–420, 429–432, 476
  whole numbers, 31–34
Resources. *See* Technology and Digital Resources
Review and Test. *See also* Assessment
  Chapter Review/Test, 55–60, 137–142, 189–194, 271–276, 317–322, 377–382, 447–452, 487–492, 539–544, 593–598, 633–638, 709–714, 749–754
  Mid-Chapter Checkpoint, 29–30, 105–106, 169–170, 233–234, 297–298, 357–358, 415–416, 467–468, 525–526, 573–574, 619–620, 671–672, 735–736
  Preview Words, 4, 62, 144, 196, 278, 326, 384, 494, 548, 600, 640, 716
  Review Words, 4, 62, 144, 196, 278, 326, 384, 454, 494, 548, 600, 640, 716
  Show What You Know, 3, 61, 143, 195, 277, 325, 383, 453, 493, 547, 599, 639, 715
Rhombus, 567–570, 581
Right angles, 549–552, 555–558, 567, 608, 723
Right triangles, 555–558
Rounding
  defined, 23
  to estimate products, 81–83, 151–154
  number line and, 23
  whole numbers, 23–26
Rulers. *See* Manipulatives and materials
Rules, 132, 173, 286–288, 306

Scale
  protractor, 613–619
Science
  Connect to Science, 26, 218, 510, 616, 688, 740
Seconds, 685–688
Sense or Nonsense?, 160, 250, 388, 411, 412, 458, 476, 504, 604, 699
Share and Show, In every Student Edition lesson. Some examples are: 7, 44, 100, 132, 172, 198, 222, 280, 329, 387, 457, 497, 551, 603, 643, 719
Show What You Know, 3, 61, 143, 195, 277, 325, 383, 453, 493, 547, 599, 639, 715
Sieve of Eratosthenes, 308
Simplest form, 339–342
Social Studies
  Connect to Social Studies, 308
Solve the Problem, 49–50, 113–114, 183–184, 265–266, 291–292, 351–352, 441–442, 481–482, 519–520, 587–588, 627–628, 691–692, 743–744
Squares, 567–570, 581, 717–719, 724–725, 729
Square units, 729–732
Standard form
  whole numbers, 11–14
Straight angle, 550–552, 608
Strategies. *See* Problem solving strategies
Student Help
  Error Alert, 38, 260, 305, 346, 582, 614, 698, 738
  Math Idea, 24, 44, 253, 279, 285, 306, 371, 575, 607, 621, 673, 724
  Read Math, 555, 685
  Remember, 39, 63, 107, 145, 158, 209, 455, 476, 513, 528, 576, 608, 723
Subtraction
  bar models, 49–52, 70–71
  draw a diagram, 49–52
  draw a model, 70–71
  equations, 49–52, 409–411
  fractions
    like denominators, 403–406, 409–412, 423–426
    modeling, 385–388, 403–406, 409–412, 429–432
  as inverse operation, 44, 50

mixed numbers, 423–426, 429–432
modeling, 49–52, 70–71
number patterns, 311–314
problem solving, 49–52
regrouping, 43–46
whole numbers, 43–46

Symmetry
lines of symmetry, 575–578
line symmetry, 581–584

Table of Measures, H38–H39
Tables
completing tables, 7, 286, 665–666, 723
making, 351–354, 648, 654, 660, 667, 686
place-value chart, 6, 17, 32–33, 495–497, 501–502, 508
tally tables, 665–667
using data from, 665–667
Venn diagrams, 556, 558, 568

Tally tables, 665–667
Technology and Digital Resources
iTools, 399, 661
Math on the Spot Videos, In every Student Edition lesson. Some examples are: 8, 46, 90, 116, 148, 217, 244, 281, 330, 388, 426, 458, 498, 552, 604, 649, 720
Multimedia eGlossary, 4, 62, 144, 196, 278, 326, 384, 454, 494, 548, 600, 640, 716
Personal Math Trainer, In some Student Edition lessons. Some examples are: 277, 294, 314, 325, 383

Tens
addition, 37–40
multiplying by, 75–78, 93–96, 99–102, 119–122, 125–128, 145–148, 163–164
place value, 5–8, 100, 221
rounding, 24–26
subtraction, 43–46

Ten thousands
addition, 37–40
place value, 6–8
rounding, 24–26
subtraction, 43–46

Tenths, 410, 495–498, 507–510, 528, 533
Term of a pattern, 311–314
Test and Review. See Review and Test

Think Smarter, In every Student Edition lesson. Some examples are: 8, 25, 63, 88, 122, 154, 185, 211, 228, 262, 281, 328, 386, 426, 457, 498, 552, 610, 644, 720
Think Smarter +, In all Student Edition chapters. Some examples are: 276, 294, 314, 388, 484, 668

Thousands
addition, 37–40
multiplying by, 75–78
place value, 5-8, 75–78
rounding, 24–26
subtraction, 43–46

Three-digit numbers
division, by one-digit divisors, 241–244, 253–256, 259–262
multiplication, three-digit by one-digit, 99–102

Time
analog clock, 601–604, 685
elapsed, 691–694

Trapezoids, 567–570, 581
Triangles
acute, 555–558
classifying by angles, 555–558
obtuse, 555–558
right, 555–558

Try Another Problem, 50, 114, 184, 266, 292, 352, 442, 482, 520, 588, 628, 692, 744
Try This!, 11, 24, 44, 64, 107, 120, 146, 152, 178, 209, 210, 286, 312, 366, 372, 436, 476, 496, 514, 527, 528, 534, 555, 561, 568, 608, 654, 698, 704, 718, 724, 738

Turns on a clock, 602
Two-digit numbers
division, by one-digit divisors, 241–244
multiplication
two-digit by one-digit, 87–90, 119–122
two-digit by two-digit, 157–160, 163–166, 171–174, 177–180

Two-dimensional figures
area
of combined rectangles, 729–732
concept of, 723–726
defined, 723
finding, 723–726, 729–732, 743–746
find unknown measures and, 737–740
formula for, 724–726
perimeter, related to, 731
of a rectangle, 723–726, 729–732
square unit, 723, 729–732

unit squares, 723
units of, 723–726
find a pattern, 587–590
hexagons, 581
identifying, 549–552, 555–558, 561–564, 567–570, 575–578, 581–584
parallelograms, 567–570, 581
perimeter
area, related to, 723–726
defined, 717
finding, 717–720
find unknown measures and, 737–740
formulas for, 717–720
measuring, 737–740
of a rectangle, 717–720
of a square, 717–720
polygons, 555–558, 567, 581
quadrilaterals, 567–570
rectangles, 567–570, 581
rhombus, 567–570, 581
shape patterns, 587–590
squares, 567–570, 581
symmetry
lines of symmetry, 575–584
line symmetry, 575–584
trapezoids, 567–570, 581
triangles, 555–558. *See also* Triangles

Understand Vocabulary, 4, 62, 144, 196, 278, 326, 384, 454, 494, 548, 600, 640, 716
Unit fractions, 391–394, 455–458
Unit squares, 723
Unlock the Problem, In every Student Edition lesson. Some examples are: 11, 49–50, 93–94, 151–152, 209–210, 279, 345, 394, 429–430, 455–456, 495–496, 533, 549–550, 578, 647–648, 717–718
Unlock the Problem Tips, 51, 115, 185, 267, 293, 353, 443, 483, 521, 589, 693, 745

Venn diagrams, 556, 558, 568, 716
Vertex, 550, 555, 575, 601, 613
Visualize It, 4, 62, 144, 196, 278, 326, 384, 454, 494, 548, 600, 640, 716

Vocabulary
Chapter Vocabulary Cards, At the beginning of every chapter.
Mid-Chapter Checkpoint, 29, 105, 169, 233, 297, 357, 415, 467, 525, 573, 619, 671, 735
Multimedia eGlossary, 4, 62, 144, 196, 278, 326, 384, 454, 494, 548, 600, 640, 716
Vocabulary, 29, 105, 233, 297, 357, 415, 467, 525, 573, 619, 671, 735
Vocabulary Builder, 4, 62, 144, 196, 278, 326, 384, 454, 494, 548, 600, 640, 716
Vocabulary Games, 4A, 62A, 144A, 196A, 278A, 326A, 384A, 454A, 494A, 548A, 600A, 640A, 716A

Week, 686–687, 703
Weight, 641, 653–656
What If, 51, 115, 185, 211, 267, 339, 353, 391, 392, 398, 443, 461, 483, 521, 522, 589, 622, 629, 648, 666, 691, 693, 697, 737, 745
What's the Error?, 13, 19, 46, 96, 134, 148, 206, 330, 409, 478, 706
What's the Question?, 20, 342, 362, 478
Whole numbers
adding, 37–40
comparing, 17–20
multiply fractions and, 469–472, 475–478
ordering, 17–20
place value, 5–8, 11–14, 17–20
renaming, 31–34
rounding, 23–26
subtracting, 43–46
Word form
whole numbers, 11–14
Write Math, In every Student Edition lesson. Some examples are: 96, 116, 224, 510, 530, 682
Writing
Write Math, In every Student Edition lesson. Some examples are: 96, 116, 224, 510, 530, 682

Yards, 641, 643, 647–650, 703

Table of Measures

METRIC

Length

1 centimeter (cm) = 10 millimeters (mm)
1 meter (m) = 1,000 millimeters
1 meter = 100 centimeters
1 meter = 10 decimeters (dm)
1 kilometer (km) = 1,000 meters

Capacity and Liquid Volume

1 liter (L) = 1,000 milliliters (mL)

Mass/Weight

1 kilogram (kg) = 1,000 grams (g)

CUSTOMARY

Length

1 foot (ft) = 12 inches (in.)
1 yard (yd) = 3 feet, or 36 inches
1 mile (mi) = 1,760 yards, or 5,280 feet

Capacity and Liquid Volume

1 cup (c) = 8 fluid ounces (fl oz)
1 pint (pt) = 2 cups
1 quart (qt) = 2 pints, or 4 cups
1 half gallon = 2 quarts
1 gallon (gal) = 2 half gallons, or 4 quarts

Mass/Weight

1 pound (lb) = 16 ounces (oz)
1 ton (T) = 2,000 pounds

TIME

1 minute (min) = 60 seconds (sec)
1 half hour = 30 minutes
1 hour (hr) = 60 minutes
1 day (d) = 24 hours
1 week (wk) = 7 days
1 year (yr) = 12 months (mo), or about 52 weeks
1 year = 365 days
1 leap year = 366 days
1 decade = 10 years
1 century = 100 years

MONEY

1 penny = 1¢, or $0.01
1 nickel = 5¢, or $0.05
1 dime = 10¢, or $0.10
1 quarter = 25¢, or $0.25
1 half dollar = 50¢, or $0.50
1 dollar = 100¢, or $1.00

SYMBOLS

<	is less than	⊥	is perpendicular to
>	is greater than	∥	is parallel to
=	is equal to	$\overleftrightarrow{AB}$	line AB
≠	is not equal to	$\overrightarrow{AB}$	ray AB
¢	cent or cents	$\overline{AB}$	line segment AB
$	dollar or dollars	∠ABC	angle ABC or angle B
°	degree or degrees	△ABC	triangle ABC

FORMULAS

	Perimeter		Area
Polygon	P = sum of the lengths of sides	Rectangle	$A = b \times h$
Rectangle	$P = (2 \times l) + (2 \times w)$ or $P = 2 \times (l + w)$		$A = l \times w$
Square	$P = 4 \times s$		